Fast Sequential Monte Carlo Methods for Counting and Optimization

Fast Sequential Monte Carlo Methods for Counting and Optimization

Reuven Y. Rubinstein

Faculty of Industrial Engineering and Management,
Technion, Israel Institute of Technology, Haifa, Israel

Ad Ridder

Department of Econometrics and Operations Research,
Vrije University, Amsterdam, Netherlands

Radislav Vaisman

Faculty of Industrial Engineering and Management,
Technion, Israel Institute of Technology, Haifa, Israel

WILEY

Cover Design: Michael Rutkowski
Cover Image: © Eastnine Inc./Getty Images

Published by John Wiley & Sons, Inc., Hoboken, New Jersey.
Published simultaneously in Canada.

Library of Congress Cataloging-in-Publication Data:

Rubinstein, Reuven Y.
Fast sequential Monte Carlo methods for counting and optimization / Reuven Rubinstein, Faculty of Industrial Engineering and Management, Technion, Israel Institute of Technology, Haifa, Israel, Ad Ridder, Department of Econometrics and Operations Research, Vrije University, Amsterdam, Netherlands, Radislav Vaisman, Faculty of Industrial Engineering and Management, Technion, Israel Institute of Technology, Haifa, Israel.
pages cm
Includes bibliographical references and index.
ISBN 978-1-118-61226-2 (cloth)–Monte Carlo method. 2. Mathematical optimization. I. Ridder, Ad, 1955- II. Vaisman, Radislav. III. Title.
T57.64.R83 2013
518′.282–dc23

2013011113

Printed in the United States of America.

10 9 8 7 6 5 4 3 2 1

In Memoriam

Reuven Y. Rubinstein (1938–2012):

Reuven Rubinstein died in December 2012, when this book was at its final stage of drafting. Reuven was an inspiration to many simulation researchers and practitioners, and was always a stimulating colleague to work with. His co-authors would like to dedicate this book to him, and to show appreciation for all the contributions he made up until the last moments of his life.

Reuven was born in Kaunas, Lithuania, in 1938. His family was exiled to the Syberian Gulag in 1941, but he managed to earn a Master's degree from Kaunas Polytechnic Institute in 1960. In 1969, Reuven obtained his PhD degree in Operational Research from Riga Polytechnical Institute. In 1973, Reuven and his family immigrated to Israel and he immediately joined the Faculty of Industrial Engineering and Management at Technion, Israel Institute of Technology. In 1978, Reuven obtained the Associate Professor rank and in 1992 he was awarded a Full Professor at Technion.

Reuven published eight books and more than one-hundred scientific papers. During his sabbatical years he was a visiting professor at many universities and research centers around the world, among them University of Illinois at Urbana-Champaign, Harvard University, Stanford University, IBM Research Center, Bell Laboratories, NJ, NEC, and the Institute of Statistical Mathematics, Japan. He was a member of several societies, including the Operations Research Society of Israel (ORSIS) and the Institute for Operations Research and Management Science (INFORMS).

A highly respected researcher in simulation algorithms, Reuven is best known for founding the score function method in simulation and the cross-entropy method for combinatorial optimization, which generated an entire new scientific community. His research was recognized with several prestigious awards, including the 2010 Lifetime Professional Achievement Award from INFORMS Simulation Society, and the 2011 Lifetime Achievement Award from ORSIS.

He will be remembered with respect and affection by everyone who loved him for his friendliness, his creative talents, and his animated spirit.

Contents

Preface

This book presents an introduction to fast sequential Monte Carlo (SMC) methods for counting and optimization. It is based mainly on the research work of Reuven Rubinstein and his collaborators, performed over the last ten years, on efficient Monte Carlo methods for estimation of rare-event probabilities, counting problems, and combinatorial optimization. Particular emphasis is placed on cross-entropy, minimum cross-entropy, splitting, and stochastic enumeration methods.

Our aim was to write a book on the SMC methods for a broad audience of engineers, computer scientists, mathematicians, statisticians, and, in general, anyone, theorist or practitioner, interested in efficient simulation and, in particular, efficient combinatorial optimization and counting. Our intention was to show how the SMC methods work in applications, while at the same time accentuating the unifying and novel mathematical ideas behind the SMC methods. We hope that the book stimulates further research at a postgraduate level.

The emphasis in this book is on concepts rather than on mathematical completeness. We assume that the reader has some basic mathematical background, such as a basic undergraduate course in probability and statistics. We have deliberately tried to avoid the formal "definition—lemma—theorem—proof" style of many mathematics books. Instead, we embed most definitions in the text and introduce and explain various concepts via examples and experiments.

Most of the combinatorial optimization and counting case studies in this book are benchmark problems taken from the World Wide Web. In all examples tested so far, the relative error of SMC was within the limits of 1–2% of the best-known solution. In some instances, SMC produced even more accurate solutions. The book covers the following topics:

Chapter 1 introduces the concepts of Monte Carlo methods for estimation and randomized algorithms to solve deterministic optimization and counting problems.

Chapter 2 deals with the cross-entropy method, which is able to approximate quite accurately the solutions to difficult estimation and optimization problems such as integer programming, continuous multiextremal optimization, noisy optimization problems such as optimal buffer allocation, optimal policy search, clustering, signal detection, DNA sequence alignment, network reliability optimization, and neural and reinforcement learning. Recently, the cross-entropy method has been used as a main engine for playing games such as Tetris, Go, and backgammon. For more references, see http://www.cemethod.org.

Chapter 3 presents the minimum cross-entropy method, also known as the MinxEnt method. Similar to the cross-entropy method, MinxEnt is able to deliver accurate solutions for difficult estimation and optimization problems. The main idea of MinxEnt is to associate with each original optimization problem an auxiliary single-constrained convex optimization program in terms of probability density functions. The beauty is that this auxiliary program has a closed-form solution, which becomes the optimal zero-variance solution, provided the "temperature" parameter is set to minus infinity. In addition, the associated probability density function based on the product of marginals obtained from the joint optimal zero variance probability density function coincides with the parametric probability density function of the cross-entropy method. Thus, we obtain a strong connection of the cross-entropy method with MinxEnt, providing solid mathematical foundations.

Chapter 4 introduces the splitting method for counting, combinatorial optimization, and rare-event estimation. Similar to the classic randomized algorithms, splitting algorithms use a sequential sampling plan to decompose a "difficult" problem into a sequence of "easy" ones. It presents, in fact, a combination of Markov chain Monte Carlo, like the Gibbs sampler, with a specially designed cloning mechanism. The latter runs in parallel multiple Markov chains by making sure that all of them run in steady-state at each iteration. This chapter contains the following elements:

- It combines splitting with the classic capture–recapture method in order to obtain low-variance estimators for counting in complex sets, such as counting the number of satisfiability assignments.
- It shows that the splitting algorithm can be efficiently used for estimating the reliability of complex static networks.
- It demonstrates numerically that the splitting algorithm can be efficiently used for generating uniform samples on discrete sets. We provide valid statistical tests supporting the uniformity of generated samples.
- It presents supportive numerical results, while solving quite general counting problems, such as counting the number of true assignments in a satisfiability problem, counting the number of feasible colorings in a graph, calculating the permanent, Hamiltonian cycles, 0-1 tables, and volume of a polytope, as well as solving integer and combinatorial optimization, like the traveling salesman, knapsack, and set-covering problems.

Chapter 5 presents a new generic sequential importance sampling algorithm, called stochastic enumeration for counting #P-complete problems, such as the number of satisfiability assignments, number of trajectories in a general network, and number of perfect matching in a graph. For this purpose, it employs a decision-making algorithm, also called an oracle. The crucial differences between stochastic enumeration and the other generic methods that we present in this book (i.e., cross-entropy, MinxEnt, and splitting) are:

- The former is sequential in nature (it is based on sequential importance sampling), whereas the latter are not (they sample simultaneously in the entire n-dimensional space).
- The former is based on decision-making oracles to solve #P-complete problems, whereas the latter are not. As a result, it is typically faster than the others.

We also present extensive numerical results with stochastic enumeration and show that it outperforms some of its well-known counterparts, in particular, the splitting method. Our explanation for that relies again on the fact that it uses sequential sampling and an oracle.

Appendix A provides auxiliary material. In particular, it covers some basic combinatorial and counting problems and supplies a background to the cross-entropy method. Finally, the efficiency of Monte Carlo estimators is discussed.

We thank all colleagues and friends who provided us with comments, corrections, and suggestions for improvements. We are grateful to the many undergraduate and graduate students at the Technion who helped make this book possible, and whose valuable ideas and experiments where extremely encouraging and motivating. We are especially indebted to Dirk Kroese, who worked through the entire manuscript and provided us a thorough review and feedback.

AD RIDDER
RADISLAV VAISMAN
Amsterdam and Haifa
September 2013

Chapter 1

Introduction to Monte Carlo Methods

Monte Carlo methods present a class of computational algorithms that rely on repeated random sampling to approximate some unknown quantities. They are best suited for calculation using a computer program, and they are typically used when the exact results with a deterministic algorithm are not available.

The Monte Carlo method was developed in the 1940s by John von Neumann, Stanislaw Ulam, and Nicholas Metropolis while they were working on the Manhattan Project at the Los Alamos National Laboratory. It was named after the Monte Carlo Casino, a famous casino where Ulam's uncle often gambled away his money.

We mainly deal in this book with two well-known Monte Carlo methods, called *importance sampling* and *splitting*, and in particular with their applications to combinatorial optimization, counting, and estimation of probabilities of rare events.

Importance sampling is a well-known variance reduction technique in stochastic simulation studies. The idea behind importance sampling is that certain values of the input random variables have a greater impact on the output parameters than others. If these "important" values are sampled more frequently, the variance of the output estimator can be reduced. However, such direct use of importance sampling distributions will result in a biased estimator. To eliminate the bias, the simulation outputs must be modified (weighted) by using a likelihood ratio factor, also called the Radon Nikodym derivative [108]. The fundamental issue in implementing importance sampling is the choice of the importance sampling distribution.

In the case of counting problems, it is well known that a straightforward application of importance sampling typically yields poor approximations of the quantity of interest. In particular, Gogate and Dechter [56, 57] show that poorly chosen importance sampling in graphical models such as satisfiability models generates many useless zero-weight samples, which are often rejected, yielding an inefficient sampling process. To address this problem, which is called the problem of

Fast Sequential Monte Carlo Methods for Counting and Optimization, First Edition.
Reuven Y. Rubinstein, Ad Ridder, and Radislav Vaisman.

losing trajectories, these authors propose a clever sample search method, which is integrated into the importance sampling framework.

With regard to probability problems, a wide range of applications of importance sampling have been reported successfully in the literature over the last decades. Siegmund [115] was the first to argue that, using an exponential change of measure, asymptotically efficient importance sampling schemes can be built for estimating gambler's ruin probabilities. His analysis is related to the theory of large deviations, which has since become an important tool for the design of efficient Monte Carlo experiments. Importance sampling is now a subject of almost any standard book on Monte Carlo simulation (see, for example, [3, 108]). We shall use importance sampling widely in this book, especially in connection to rare-event estimation.

The splitting method dates back to Kahn and Harris [62] and Rosenbluth and Rosenbluth [97]. The main idea is to partition the state-space of a system into a series of nested subsets and to consider the rare event as the intersection of a nested sequence of events. When a given subset is entered by a sample trajectory during the simulation, numerous random retrials are generated, with the initial state for each retrial being the state of the system at the entry point. By doing so, the system trajectory is split into a number of new subtrajectories, hence the name "splitting". Since then, hundreds of papers have been written on this topic, both from a theoretical and a practical point of view. Applications of the splitting method arise in particle transmission (Kahn and Harris [62]), queueing systems (Garvels [48], Garvels and Kroese [49], Garvels et al. [50]), and reliability (L'Ecuyer et al. [76]). The method has been given new impetus by the RESTART (Repetitive Simulation Trials After Reaching Thresholds) method in the sequence of papers by Villén-Altimirano and Villén-Altimirano [122–124]. A fundamental theory of the splitting method was developed by Melas [85], Glasserman et al. [54, 55], and Dean and Dupuis [38, 39]. Recent developments include the adaptive selection of the splitting levels in Cérou and Guyader [24], the use of splitting in reliability networks [73, 109], quasi-Monte Carlo estimators in L'Ecuyer et al. [77], and the connection between splitting for Markovian processes and interacting particle methods based on the Feynman-Kac model in Del Moral [89].

Let us introduce the notion of a *randomized algorithm*. A randomized algorithm is an algorithm that employs a degree of randomness as part of its logic to solve a deterministic problem such as a combinatorial optimization problem. As a result, the algorithm's output will be a random variable representing either the running time, its output, or both. In general, introducing randomness may result in an algorithm that is far simpler, more elegant, and sometimes even more efficient than the deterministic counterpart.

EXAMPLE 1.1 ***Checking Matrix Multiplication***

Suppose we are given three $n \times n$ matrices $\boldsymbol{A}$, $\boldsymbol{B}$, and $\boldsymbol{C}$ and we want to check whether $\boldsymbol{AB} = \boldsymbol{C}$.

A trivial deterministic algorithm would be to run a standard multiplication algorithm and compare each entry of $\boldsymbol{AB}$ with $\boldsymbol{C}$. Simple matrix multiplication requires $\mathcal{O}(n^3)$ operations. A more sophisticated algorithm [88] takes only $\mathcal{O}(n^{2.376})$ operations. Using randomization, however, we need only $\mathcal{O}(n^2)$ operations, with an extremely small probability of error [88].

The randomized procedure is as follows:

- Pick a random n-dimensional vector $\boldsymbol{r} = (r_1, \ldots, r_n)$.
- Multiply both sides of $\boldsymbol{AB} = \boldsymbol{C}$ by $\boldsymbol{r}$, that is, obtain $\boldsymbol{A}(\boldsymbol{Br})$ and $\boldsymbol{Cr}$.
- If $\boldsymbol{A}(\boldsymbol{Br}) = \boldsymbol{Cr}$, then declare $\boldsymbol{AB} = \boldsymbol{C}$, otherwise, $\boldsymbol{AB} \neq \boldsymbol{C}$.

This algorithm runs in $\mathcal{O}(n^2)$ operations because matrix multiplication is associative, so $(\boldsymbol{AB})\boldsymbol{r}$ can be computed as $\boldsymbol{A}(\boldsymbol{Br})$, thus requiring only three matrix-vector multiplications for the algorithm. □

For more examples and foundations on randomized algorithms, see the monographs [88, 90].

We shall consider not only randomized algorithms but also random structures. The latter comprises random graphs (such as Erdös-Rényi graphs), random Boolean formulas, and so on. Random structures are of interest both as a means of understanding the behavior of algorithms on typical inputs and as a mathematical framework in which one can investigate various probabilistic techniques to analyze randomized algorithms.

This book deals with Monte Carlo methods and their associated randomized algorithms for solving combinatorial optimization and counting problems. In particular, we consider combinatorial problems that can be modeled by integer linear constraints. To clarify, denote by $\mathcal{X}^*$ the set of feasible solutions of a combinatorial problem, which is assumed to be a subset of an n-dimensional integer vector space and which is given by the following linear constraints:

$$\mathcal{X}^* = \begin{cases} \sum_{j=1}^{n} a_{ij} x_j = b_i, & \text{for all } \ i = 1, \ldots, m_1 \\ \sum_{j=1}^{n} a_{ij} x_j \geq b_i, & \text{for all } \ i = m_1 + 1, \ldots, m_1 + m_2 = m \\ \boldsymbol{x} = (x_1, \ldots, x_n) \in \mathbb{Z}^n. & \end{cases} \tag{1.1}$$

Here, $\boldsymbol{A} = (a_{ij})$ is a given $m \times n$ matrix and $\boldsymbol{b} = (b_i)$ is a given m-dimensional vector. Most often we require the variables x_j to be nonnegative integers and, in particular, binary integers.

In this book, we describe in detail various problems, algorithms, and mathematical aspects that are associated with (1.1) and its relation to decision making, counting, and optimization. Below is a short list of problems associated with (1.1):

1. *Decision making*: Is $\mathcal{X}^*$ nonempty?
2. *Optimization*: Solve $\max_{\boldsymbol{x} \in \mathcal{X}^*} S(\boldsymbol{x})$ for a given objective (performance) function $S : \mathcal{X}^* \to \mathbb{R}$.
3. *Counting*: Calculate the cardinality $|\mathcal{X}^*|$ of $\mathcal{X}^*$.

It turns out that, typically, it is hard to solve any of the above three problems and, in particular, the counting one, which is the hardest one. However, we would like to point out that there are problems for which decision making is easy (polynomial time) but counting is hard [90]. As an example, finding a feasible path (and also the shortest path) between two fixed nodes in a network is easy, whereas counting the total number of paths between the two nodes is difficult. Some other examples of hard counting and easy decision-making problems include:

- How many different variable assignments will satisfy a given satisfiability formula in disjunctive normal form?
- How many different variable assignments will satisfy a given 2-satisfiability formula (constraints on pairs of variables)?
- How many perfect matchings are there for a given bipartite graph?

In Chapter 5, we follow the saying "counting is hard, but decision making is easy" and employ relevant decision-making algorithms, also called oracles, to derive fast Monte Carlo algorithms for counting.

Below is a detailed list of interesting hard counting problems.

- *The Hamiltonian cycle problem.* How many Hamiltonian cycles does a graph have? That is, how many tours contains a graph in which every node is visited exactly once (except for the beginning/end node)?
- *The permanent problem.* Calculate the permanent of a matrix $\boldsymbol{A}$, or equivalently, the number of perfect matchings in a bipartite balanced graph with $\boldsymbol{A}$ as its biadjacency matrix.
- *The self-avoiding walk problem.* How many self-avoiding random walks of length n exist, when we are allowed to move at each grid point in any neighboring direction with equal probability?
- *The connectivity problem.* Given two different nodes in a directed or undirected graph, say v and w, how many paths exist from v to w that do not traverse the same edge more than once?
- *The satisfiability problem.* Let $\mathcal{X}$ be a collection of all sets of n Boolean variables $\{x_1, \ldots, x_n\}$. Thus, $\mathcal{X}$ has cardinality $|\mathcal{X}| = 2^n$. Let $\mathcal{C}$ be a set of m Boolean disjunctive clauses. Examples of such clauses are $C_1 = x_1 \vee \bar{x}_2 \vee x_4$, $C_2 = \bar{x}_2 \vee \bar{x}_3$, etc. How many (if any) satisfying truth assignments for $\mathcal{C}$ exist, that is, how many ways are there to set the variables $x_1, \ldots, x_n$ either true or false so that all clauses $C_i \in \mathcal{C}$ are true?
- *The k-coloring problem.* Given $k \geq 3$ distinct colors, in how many different ways can one color the nodes (or the edges) of a graph, so that each two adjacent nodes (edges, respectively) in the graph have different colors?
- *The spanning tree problem.* How many unlabeled spanning trees has a graph G? Note that this counting problem is easy for labeled graphs.
- *The isomorphism problem.* How many isomorphisms exist between two given graphs G and H? In other words, in an isomorphism problem one needs to

find all mappings ϕ between the nodes of G and H such that (v, w) is an edge of G if and only if $\phi(v)\phi(w)$ is an edge of H.

- *The clique problem.* How many cliques of fixed size k exist in a graph G? Recall that a clique is a complete subgraph of G.

The decision versions of these problems are all examples of NP-complete problems. Clearly, counting all feasible solutions, denoted by #P, is an even harder problem.

Generally, the complexity class #P consists of the counting problems associated with the decision problems in NP. Completeness is defined similarly to the decision problems: a problem is #P-complete if it is in #P, and if every #P problem can be reduced to it in polynomial counting reduction. Hence, the counting problems that we presented above are all #P-complete. For more details we refer the reader to the classic monograph by Papadimitrou and Stieglitz [92].

Chapter 2

Cross-Entropy Method

2.1 INTRODUCTION

The cross-entropy (CE) method is a powerful technique for solving difficult estimation and optimization problems, based on Kullback-Leibler (or cross-entropy) minimization [15]. It was pioneered by Rubinstein in 1999 [100] as an adaptive importance sampling procedure for the estimation of rare-event probabilities. Subsequent work in [101, 102] has shown that many optimization problems can be translated into a rare-event estimation problem. As a result, the CE method can be utilized as randomized algorithms for optimization. The gist of the idea is that the probability of locating an optimal or near-optimal solution using naive random search is a rare event probability. The cross-entropy method can be used to gradually change the sampling distribution of the random search so that the rare event is more likely to occur. For this purpose, using the cross-entropy or Kullback-Leibler divergence as a measure of closeness between two sampling distributions, the method estimates a sequence of parametric sampling distributions that converges to a distribution with probability mass concentrated in a region of near-optimal solutions.

To date, the CE method has been successfully applied to optimization and estimation problems. The former includes mixed-integer nonlinear programming [69], continuous optimal control problems [111, 112], continuous multiextremal optimization [71], multidimensional independent component analysis [116], optimal policy search [19], clustering [11, 17, 72], signal detection [80], DNA sequence alignment [67, 93], fiber design [27], noisy optimization problems such as optimal buffer allocation [2], resource allocation in stochastic systems [31], network reliability optimization [70], vehicle routing optimization with stochastic demands [29], power system combinatorial optimization problems [45], and neural and reinforcement learning [81, 86, 118, 128]. CE has even been used as a main engine for playing games such as Tetris, Go, and backgammon [26].

In the estimation setting, the CE method can be viewed as an adaptive importance sampling procedure. In the optimization setting, the optimization

Fast Sequential Monte Carlo Methods for Counting and Optimization, First Edition.
Reuven Y. Rubinstein, Ad Ridder, and Radislav Vaisman.

problem is first translated into a rare-event estimation problem, and then the CE method for estimation is used as an adaptive algorithm to locate the optimum.

The CE method is based on a simple iterative procedure where each iteration contains two phases: (a) generate a random data sample (trajectories, vectors, etc.) according to a specified mechanism; (b) update the parameters of the random mechanism on the basis of the data, in order to produce a "better" sample in the next iteration.

Among the two, optimization and estimation, CE is best known for optimization, for which it is able to handle problems of size up to thousands in a reasonable time limit. For estimation and also counting problems, multilevel CE can be useful only for problems of small and moderate size. The main reason for that is, unlike optimization, in estimation multilevel CE involves likelihood ratios factors for all variables. The well-known degeneracy of the likelihood ratio estimators [106] causes CE to perform poorly in high dimensions. To overcome this problem, [25] proposes a variation of the CE method by considering a single iteration for determining the importance sampling density. For this approach to work, one must be able to generate samples from the optimal (zero-variance) density.

A tutorial on the CE method is given in [37]. A comprehensive treatment can be found in [107]. The CE method website (www.cemethod.org) lists several hundred publications.

The rest of this chapter is organized as follows. Section 2.2 presents a general CE algorithm for the estimation of rare event probabilities, while Section 2.3 introduces a slight modification of this algorithm for solving combinatorial optimization problems. We consider applications of the CE method to the multidimensional knapsack problem, to the Mastermind game, and to Markov decision problems. Also, we provide supportive numerical results on the performance of the algorithm for these problems. Sections 2.4 and 2.5 discuss advanced versions of CE to deal with continuous and noisy optimization, respectively.

2.2 ESTIMATION OF RARE-EVENT PROBABILITIES

Consider the problem of computing a probability of the form

$$\ell = \mathbb{P}_h(S(\boldsymbol{X}) \geq \gamma) = \mathbb{E}_h[I_{\{S(\boldsymbol{X}) \geq \gamma\}}], \tag{2.1}$$

for some fixed level γ. In this expression $S(\boldsymbol{X})$ is the sample performance, and $\boldsymbol{X}$ a random vector with probability density function (pdf) $h(\cdot)$, sometimes called the prior pdf. The standard or crude Monte Carlo estimation method is based on repetitive sampling from h and using the sample average:

$$\hat{\ell} = \frac{1}{N} \sum_{k=1}^{N} I_{\{S(\boldsymbol{X}_k) \geq \gamma\}}, \tag{2.2}$$

where $\boldsymbol{X}_1, \ldots, \boldsymbol{X}_N$ are independent and identically distributed (iid) samples drawn from h. However, for rare-event probabilities, say $\ell \approx 10^{-9}$ or less, this method fails for the following reasons [108]. The efficiency of the estimator is measured by its relative error (RE), which is defined as

$$\mathrm{RE}(\hat{\ell}) = \frac{\sqrt{\mathbb{V}\mathrm{ar}[\hat{\ell}]}}{\mathbb{E}[\hat{\ell}]}. \tag{2.3}$$

Clearly, for the standard estimator (2.2)

$$\mathbb{V}\mathrm{ar}[\hat{\ell}] = \frac{1}{N}\mathbb{V}\mathrm{ar}[I_{\{S(\boldsymbol{X})\geq\gamma\}}] = \frac{1}{N}\ell(1-\ell).$$

Thus, for small probabilities, meaning $1-\ell \approx 1$, we get

$$\mathrm{RE}(\hat{\ell}) = \frac{\sqrt{1-\ell}}{\sqrt{N\ell}} \approx \frac{1}{\sqrt{N\ell}} = \sqrt{\frac{\ell^{-1}}{N}}.$$

This says that the sample size N is of the order ℓ^{-1}/ε^2 to obtain relative error of ε. For rare-event probabilities, this sample size leads to unmanageable computation time. For instance, let $S(\boldsymbol{X})$ be the maximum number of customers that during a busy period of an $M/M/1$ queue is present. For a load at 80%, the probability can be computed to be $\ell = \mathbb{P}(S(\boldsymbol{X}) \geq 100) \approx 4 \cdot 10^{-11}$. A simulation to obtain relative error of 10% would require about $2.5 \cdot 10^{12}$ samples of $\boldsymbol{X}$ (busy cycles). On a 2.79 Ghz (0.99GB RAM) PC this would take about 18 days.

Assuming asymptotic normality of the estimator, the confidence intervals (CI) follow in a standard way. For example, a 95% CI for ℓ is given by

$$\widehat{\ell} \pm 1.96\, \widehat{\ell}\,\mathrm{RE}.$$

This shows again the reciprocal relationship between sample size and rare-event probability to obtain accuracy: suppose one wishes that the confidence half width is at most $100\varepsilon\%$ relative to the estimate $\widehat{\ell}$. This boils down to

$$\frac{1.96\widehat{\ell}\,\mathrm{RE}}{\widehat{\ell}} < \varepsilon \quad \Leftrightarrow \quad 1.96\,\mathrm{RE} < \varepsilon.$$

For variance reduction we apply *importance sampling* simulation using pdf $g(\cdot)$, that is, we draw independent and identically distributed samples $\boldsymbol{X}_1, \ldots, \boldsymbol{X}_N$ from g, calculate their *likelihood ratios*

$$W(\boldsymbol{X}_k; h, g) = h(\boldsymbol{X}_k)/g(\boldsymbol{X}_k),$$

and set the importance sampling estimator by

$$\widehat{\ell} = \frac{1}{N}\sum_{k=1}^{N} I_{\{S(\boldsymbol{X}_k)\geq\gamma\}}\; W(\boldsymbol{X}_k; h, g). \tag{2.4}$$

By writing $Z_k = I_{\{S(\boldsymbol{X}_k)\geq\gamma\}}\, W(\boldsymbol{X}_k; h, g)$, we see that this estimator is of the form $\frac{1}{N}\sum_{k=1}^{N} Z_k$. Again, we measure efficiency by its relative error, similar to (2.3), which might be estimated by $S/(\widehat{\ell}\sqrt{N})$, with

$$S^2 = \frac{1}{N-1}\sum_{k=1}^{N}(Z_k - \widehat{\ell})^2. \qquad (2.5)$$

being the sample variance of the $\{Z_k\}$.

In almost all realistic problems, the prior pdf h belongs to a parameterized family

$$\mathscr{F} = \{f(\cdot; \boldsymbol{v}) : \boldsymbol{v} \in \mathcal{V}\}. \qquad (2.6)$$

That means that $h(\cdot) = f(\cdot; \boldsymbol{u})$ is given by some fixed parameter $\boldsymbol{u} \in \mathcal{V}$. Then it is natural to choose the importance sampling pdf $g(\cdot) = f(\cdot; \boldsymbol{v})$ from the same family using another parameter $\boldsymbol{v}$, or $\boldsymbol{w}$, etc. In that case the likelihood ratio becomes

$$W(\boldsymbol{X}_k; \boldsymbol{u}, \boldsymbol{v}) = \frac{f(\boldsymbol{X}_k; \boldsymbol{u})}{f(\boldsymbol{X}_k; \boldsymbol{v})}, \qquad (2.7)$$

We refer to the original parameter $\boldsymbol{u}$ as the *nominal parameter* of the problem, and we call parameters $\boldsymbol{v} \in \mathcal{V}$ used in the importance sampling, the *reference parameters*.

EXAMPLE 2.1 ***Stochastic Shortest Path***

Our objective is to efficiently estimate the probability ℓ that the shortest path from node s to node t in the network of Figure 2.1 has a length of at least γ. The random lengths $X_1, \ldots, X_5$ of the links are assumed to be independent and exponentially distributed with means $u_1, \ldots, u_5$, respectively.

Defining $\boldsymbol{X} = (X_1, \ldots, X_5)$, $\boldsymbol{u} = (u_1, \ldots, u_n)$, and

$$S(\boldsymbol{X}) = \min\{X_1 + X_4,\ X_1 + X_3 + X_5,\ X_2 + X_5,\ X_2 + X_3 + X_4\},$$

the problem is cast in the framework of (2.1). Note that we can estimate (2.1) via (2.4) by drawing $X_1, \ldots, X_5$ independently from exponential distributions that are possibly different from the original ones. That is, $X_i \sim \mathsf{Exp}(v_i^{-1})$ instead of $X_i \sim \mathsf{Exp}(u_i^{-1})$, $i = 1, \ldots, 5$.

The challenging problem is how to select a vector $\boldsymbol{v} = (v_1, \ldots, v_5)$ that gives the most accurate estimate of ℓ for a given simulation effort. To do so we use a two-stage procedure, where both the level γ and the reference parameters $\boldsymbol{v}$ are updated simultaneously. As we shall soon see, one of the strengths of the CE method for rare-event simulation is that it provides a fast way to estimate accurately the optimal parameter vector $\boldsymbol{v}^*$. □

The idea of the CE method is to choose the importance sampling pdf g from within the parametric class of pdfs $\mathscr{F}$ such that the Kullback-Leibler entropy

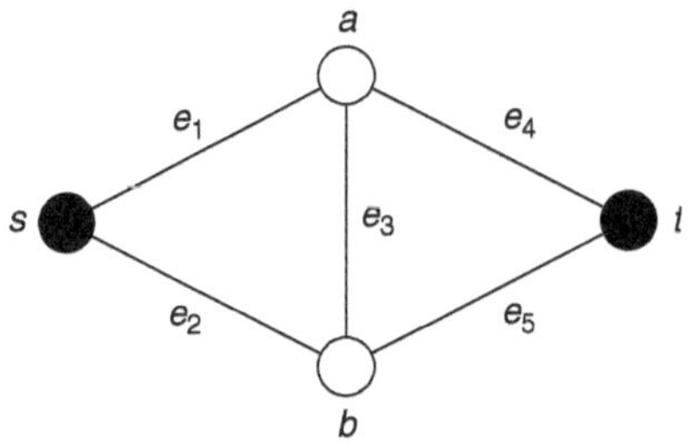

Figure 2.1 Shortest path from s to t.

between g and the optimal (zero variance) importance sampling pdf g^*, given by [107],

$$g^*(\boldsymbol{x}) = \frac{f(\boldsymbol{x};\boldsymbol{u})\, I_{\{S(\boldsymbol{x})\geq\gamma\}}}{\ell},$$

is minimal. The Kullback-Leibler divergence between g^* and g is given by

$$\mathcal{D}(g^*, g) = \mathbb{E}_{g^*}\left[\log\frac{g^*(\boldsymbol{X})}{g(\boldsymbol{X})}\right] = \int g^*(\boldsymbol{x})\log\frac{g^*(\boldsymbol{x})}{g(\boldsymbol{x})}\,\mathrm{d}\boldsymbol{x}. \tag{2.8}$$

The CE minimization procedure then reduces to finding an optimal reference parameter, $\boldsymbol{v}^*$, say by cross-entropy minimization:

$$\begin{aligned}\boldsymbol{v}^* &= \underset{\boldsymbol{v}}{\operatorname{argmin}}\ \mathcal{D}(g^*, f(\cdot;\boldsymbol{v}))\\ &= \underset{\boldsymbol{v}}{\operatorname{argmax}}\ \mathbb{E}_{\boldsymbol{u}}[I_{\{S(\boldsymbol{X})\geq\gamma\}}\log f(\boldsymbol{X};\boldsymbol{v})]\\ &= \underset{\boldsymbol{v}}{\operatorname{argmax}}\ \mathbb{E}_{\boldsymbol{w}}[I_{\{S(\boldsymbol{X})\geq\gamma\}}\log f(\boldsymbol{X};\boldsymbol{v})W(\boldsymbol{X},\boldsymbol{u},\boldsymbol{w})],\end{aligned} \tag{2.9}$$

where $\boldsymbol{w}$ is any reference parameter and the subscript in the expectation operator indicates the density of $\boldsymbol{X}$. This $\boldsymbol{v}^*$ can be estimated via the stochastic counterpart of (2.9):

$$\widehat{\boldsymbol{v}} = \underset{\boldsymbol{v}}{\operatorname{argmax}}\frac{1}{N}\sum_{k=1}^{N} I_{\{S(\boldsymbol{X})\geq\gamma\}}W(\boldsymbol{X}_k,\boldsymbol{u},\boldsymbol{w})\ \log f(\boldsymbol{X}_k;\boldsymbol{v}), \tag{2.10}$$

where $\boldsymbol{X}_1,\ldots,\boldsymbol{X}_N$ are drawn from $f(\cdot;\boldsymbol{w})$.

The optimal parameter $\widehat{\boldsymbol{v}}$ in (2.10) can often be obtained in explicit form, in particular when the class of sampling distributions forms an exponential family [107, pages 319–320]. Indeed, analytical updating formulas can be found whenever explicit expressions for the maximal likelihood estimators of the parameters can be found [37, page 36].

In the other cases, we consider a multilevel approach, where we generate a sequence of reference parameters $\{\boldsymbol{v}_t,\ t\geq 0\}$ and a sequence of levels $\{\gamma_t,\ t\geq 1\}$, while iterating in both γ_t and $\boldsymbol{v}_t$. Our ultimate goal is to have $\boldsymbol{v}_t$ close to $\boldsymbol{v}^*$ after some number of iterations and to use $\boldsymbol{v}_t$ in the importance sampling density $f(\cdot;\boldsymbol{v}_t)$ to estimate ℓ.

We start with $\boldsymbol{v}_0 = \boldsymbol{u}$ and choose ρ to be a not too small number, say $10^{-2} \leq \rho \leq 10^{-1}$. At the first iteration, we choose $\boldsymbol{v}_1$ to be the optimal parameter for estimating $\mathbb{P}_{\boldsymbol{u}}(S(\boldsymbol{X}) \geq \gamma_1)$, where γ_1 is the $(1-\rho)$-quantile of $S(\boldsymbol{X})$. That is, γ_1 is the largest real number for which $\mathbb{P}_{\boldsymbol{u}}(S(\boldsymbol{X}) \geq \gamma_1) \geq \rho$. Thus, if we simulate under $\boldsymbol{u}$, then level γ_1 is reached with a reasonably high probability of around ρ. In this way we can estimate the (unknown!) level γ_1, simply by generating a random sample $\boldsymbol{X}_1, \ldots, \boldsymbol{X}_N$ from $f(\bullet; \boldsymbol{u})$, and calculating the $(1-\rho)$-quantile of the order statistics of the performances $S(\boldsymbol{X}_i)$; that is, denote the ordered set of performance values by $S_{(1)}, S_{(2)}, \ldots, S_{(N)}$, then

$$\widehat{\gamma}_1 = S_{(\lceil (1-\rho)N \rceil)},$$

where $\lceil \quad \rceil$ denotes the integer part. Next, the reference parameter $\boldsymbol{v}_1$ can be estimated via (2.10), replacing γ with the estimate of γ_1. Note that we can use here the same sample for estimating both $\boldsymbol{v}_1$ and γ_1. This means that $\boldsymbol{v}_1$ is estimated on the basis of the $\lceil (1-\rho)N \rceil$ best samples, that is, the samples $\boldsymbol{X}_i$ for which $S(\boldsymbol{X}_i)$ is greater than or equal to $\widehat{\gamma}_1$. These form the elite samples in the first iteration. We repeat these steps in the subsequent iterations. The two updating phases, starting from $\boldsymbol{v}_0 = \widehat{\boldsymbol{v}}_0 = \boldsymbol{u}$ are given below:

1. **Adaptive updating of $\boldsymbol{\gamma_t}$.** For a fixed $\boldsymbol{v}_{t-1}$, let γ_t be the $(1-\rho)$-quantile of $S(\boldsymbol{X})$ under $\boldsymbol{v}_{t-1}$. To estimate γ_t, draw a random sample $\boldsymbol{X}_1, \ldots, \boldsymbol{X}_N$ from $f(\bullet; \widehat{\boldsymbol{v}}_{t-1})$ and evaluate the sample $(1-\rho)$-quantile $\widehat{\gamma}_t$.
2. **Adaptive updating of $\boldsymbol{v}_t$.** For fixed γ_t and $\boldsymbol{v}_{t-1}$, derive $\boldsymbol{v}_t$ as

$$\boldsymbol{v}_t = \underset{\boldsymbol{v} \in \mathcal{V}}{\operatorname{argmax}} \, \mathbb{E}_{\boldsymbol{v}_{t-1}} [I_{\{S(\boldsymbol{X}) \geq \gamma_t\}} W(\boldsymbol{X}; \boldsymbol{u}, \boldsymbol{v}_{t-1}) \log f(\boldsymbol{X}; \boldsymbol{v})]. \tag{2.11}$$

The stochastic counterpart of (2.11) is as follows: for fixed $\widehat{\gamma}_t$ and $\widehat{\boldsymbol{v}}_{t-1}$, derive $\widehat{\boldsymbol{v}}_t$ as the solution

$$\widehat{\boldsymbol{v}}_t = \underset{\boldsymbol{v} \in \mathcal{V}}{\operatorname{argmax}} \, \frac{1}{N} \sum_{\boldsymbol{X}_k \in \mathcal{E}_t} W(\boldsymbol{X}_k; \boldsymbol{u}, \widehat{\boldsymbol{v}}_{t-1}) \log f(\boldsymbol{X}_k; \boldsymbol{v}), \tag{2.12}$$

where $\mathcal{E}_t$ is the set of elite samples in the t-th iteration, that is, the samples $\boldsymbol{X}_k$ for which $S(\boldsymbol{X}_k) \geq \widehat{\gamma}_t$.

The procedure terminates when at some iteration T a level $\widehat{\gamma}_T$ is reached that is at least γ, and thus the original value of γ can be used without getting too few samples. We then reset $\widehat{\gamma}_T$ to γ, reset the corresponding elite set, and deliver the final reference parameter $\widehat{\boldsymbol{v}}^*$, using again (2.12). This $\widehat{\boldsymbol{v}}^*$ is then used in (2.4) to estimate ℓ.

Remark 2.1 ***Improved Cross-Entropy Method***

The multilevel approach may not be appropriate for problems that cannot cast in the framework of (2.1). Specifically, when we deal with a high-dimensional problem there is often not a natural modeling in terms of level crossings. Furthermore, in high-dimensional settings, the likelihood ratio degeneracy becomes a severe issue causing the importance sampling estimator to be unreliable [106]. Recent enhancements of the cross-entropy method have been developed to overcome this problem. A successful approach has been proposed in [25]. It is based on applying a Markov chain Monte Carlo method to generate samples, of the zero-variance density. With these samples, one maximizes the logarithm of the proposed density with respect to the reference parameter $\boldsymbol{v}$. The advantage is that it does not involve the likelihood ratio in this optimization program.

Remark 2.2 ***Smoothed Updating***

Instead of updating the parameter vector $\boldsymbol{v}$ directly via (2.12), we use the following smoothed version:

$$\widehat{\boldsymbol{v}}_t = \alpha \tilde{\boldsymbol{v}}_t + (1-\alpha)\widehat{\boldsymbol{v}}_{t-1}, \tag{2.13}$$

where $\tilde{\boldsymbol{v}}_t$ is the parameter vector given in (2.12), and α is called the smoothing parameter, typically with $0.7 < \alpha \leq 1$. Clearly, for $\alpha = 1$ we have our original updating rule. The reason for using the smoothed (2.13) instead of the original updating rule is twofold: (a) to smooth out the values of $\widehat{\boldsymbol{v}}_t$, (b) to reduce the probability that some component $\widehat{v}_{t,i}$ of $\widehat{\boldsymbol{v}}_t$ will be zero or one at the first few iterations. This is particularly important when $\widehat{\boldsymbol{v}}_t$ is a vector or matrix of probabilities. Because in those cases, when $0 < \alpha \leq 1$ it is always ensured that $\widehat{v}_{t,i} > 0$, while for $\alpha = 1$ one might have (even at the first iterations) that either $\widehat{v}_{t,i} = 0$ or $\widehat{v}_{t,i} = 1$ for some indices i. As a result, the algorithm will converge to a wrong solution.

The resulting CE algorithm for rare-event probability estimation can thus be written as follows.

Algorithm 2.1 ***CE Algorithm for Rare-Event Estimation***

1. ***Initialize:*** *Define* $\widehat{\boldsymbol{v}}_0 = \boldsymbol{u}$. *Let* $N^{elite} = \lceil (1-\rho)N \rceil$. *Set* $t = 1$ *(iteration counter).*
2. ***Draw:*** *Generate a random sample* $\boldsymbol{X}_1, \ldots, \boldsymbol{X}_N$ *according to the pdf* $f(\cdot; \widehat{\boldsymbol{v}}_{t-1})$.
3. ***Select:*** *Calculate the performances* $S(\boldsymbol{X}_k)$ *for all k, and order them from smallest to biggest,* $S_{(1)} \leq \ldots \leq S_{(N)}$. *Let* $\widehat{\gamma}_t$ *be the sample* $(1-\rho)$*-quantile of performances:* $\widehat{\gamma}_t = S_{(N^{elite})}$. *If* $\widehat{\gamma}_t > \gamma$ *reset* $\widehat{\gamma}_t$ *to* γ.
4. ***Update:*** *Use the same sample* $\boldsymbol{X}_1, \ldots, \boldsymbol{X}_N$ *to solve the stochastic program* (2.12).
5. ***Smooth:*** *Apply* (2.13) *to smooth out the vector* $\tilde{\boldsymbol{v}}_t$.
6. ***Iterate:*** *If* $\widehat{\gamma}_t < \gamma$, *set* $t = t + 1$ *and reiterate from step 2. Else proceed with step 7.*
7. ***Final:*** *Let* T *be the final iteration counter. Generate a sample* $\boldsymbol{X}_1, \ldots, \boldsymbol{X}_{N_1}$ *according to the pdf* $f(\cdot; \widehat{\boldsymbol{v}}_T)$ *and estimate* ℓ *via* (2.4).

Apart from specifying the family of sampling pdfs, the sample sizes N and N_1, and the rarity parameter ρ, the algorithm is completely self-tuning. The sample size N for determining a good reference parameter can usually be chosen much smaller than the sample size N_1 for the final importance sampling estimation, say $N = 1000$ versus $N_1 = 100{,}000$. Under certain technical conditions [107], the deterministic version of Algorithm 2.1 is guaranteed to terminate (reach level γ) provided that ρ is chosen small enough.

Note that Algorithm 2.1 breaks down the "hard" problem of estimating the very small probability ℓ into a sequence of "simple" problems, generating a sequence of pairs $\{(\widehat{\gamma}_t, \widehat{\boldsymbol{v}}_t)\}$, depending on ρ, which is called the rarity parameter. Convergence of Algorithm 2.1 is discussed in [107]. Other convergence proofs for the CE method may be found in [33] and [84].

Remark 2.3 ***Maximum Likelihood Estimator***

Optimization problems of the form (2.12) appear frequently in statistics. In particular, if the W factor is omitted, which will turn out to be important in CE optimization, one can write (2.12) also as

$$\widehat{\boldsymbol{v}}_t = \underset{\boldsymbol{v}}{\operatorname{argmax}} \prod_{\boldsymbol{X}_k \in \mathcal{E}_t} f(\boldsymbol{X}_k; \boldsymbol{v}),$$

where the product is the joint density of the elite samples. Consequently, $\widehat{\boldsymbol{v}}_t$ is chosen such that the joint density of the elite samples is maximized. Viewed as a function of the parameter $\boldsymbol{v}$, rather than the data $\{\mathcal{E}_t\}$, this function is called the likelihood. In other words, $\widehat{\boldsymbol{v}}_t$ is the maximum likelihood estimator (it maximizes the likelihood) of $\boldsymbol{v}$ based on the elite samples. When the W factor is present, the form of the updating formula remains similar.

To provide further insight into Algorithm 2.1, we shall follow it step-by-step in a couple of toy examples.

EXAMPLE 2.2 ***Binomial Distribution***

Assume that we want to estimate via simulation the probability $\ell = \mathbb{P}_u(X \geq \gamma)$, where $X \sim \mathsf{Bin}(n, u)$, $u \in (0, 1)$ is the success probability, and $\gamma \in \{0, 1, \ldots, n\}$ is the target threshold. Suppose that γ is large compared to the mean $\mathbb{E}[X] = nu$, so that

$$\ell = \sum_{m=\gamma}^{n} \binom{n}{m} u^m (1 - u)^{n-m}$$

is a rare-event probability. Assume further that n is fixed and we want to estimate the optimal parameter $\widehat{v}_t \in (0, 1)$ in $\mathsf{Bin}(n, v)$ using (2.12).

In iteration t, we sample independent and identically distributed $X_1, \ldots, X_N \sim \mathsf{Bin}(n, \widehat{v}_{t-1})$. The right-hand side of (2.12) becomes

$$\sum_{X_k \in \mathcal{E}_t} W_k \log\left(\binom{n}{X_k} v^{X_k} (1 - v)^{n - X_k}\right)$$

Table 2.1 The evolution of $\widehat{\gamma}_t$ and $\widehat{v}_t$ for $\gamma = 50$, using $\rho = 0.05$ and $N = 1000$ samples

t	$\widehat{\gamma}_t$	$\widehat{v}_t$
0	—	0.1667
1	23	0.2530
2	33	0.3376
3	42	0.4263
4	51	0.5124
5	50	0.5028

$$= \sum_{X_k \in \mathcal{E}_t} W_k \log \binom{n}{X_k} + \sum_{X_k \in \mathcal{E}_t} W_k X_k \log v + \sum_{X_k \in \mathcal{E}_t} W_k (n - X_k) \log(1 - v),$$

where

$$W_k = W(X_k; u, \widehat{v}_{t-1}) = \left(\frac{u}{\widehat{v}_{t-1}}\right)^{X_k} \left(\frac{1-u}{1-\widehat{v}_{t-1}}\right)^{n-X_k}$$

is the likelihood ratio of the k-th sample. To find the optimal $\widehat{v}_t$, we consider the first-order condition. We obtain

$$\sum_{X_k \in \mathcal{E}_t} W_k X_k \frac{1}{v} - \sum_{X_k \in \mathcal{E}_t} W_k (n - X_k) \frac{1}{1-v} = 0.$$

Solving this for v yields

$$\widehat{v}_t = \frac{\sum_{X_k \in \mathcal{E}_t} W_k \, X_k / n}{\sum_{X_k \in \mathcal{E}_t} W_k}. \tag{2.14}$$

In other words, $\widehat{v}_t$ is simply the sample fraction of the elite samples, weighted by the likelihood ratios. Note that, without the weights $\{W_k\}$, we would obtain the maximum likelihood estimator of v for the $\mathsf{Bin}(n, v)$ distribution based on the elite samples, in accordance with Remark 2.3. Note also that (2.14) can be viewed as a stochastic counterpart of the deterministic updating formula

$$v_t = \frac{\mathbb{E}_u[I_{\{X \geq \gamma_t\}} \, X/n]}{\mathbb{E}_u[I_{\{X \geq \gamma_t\}}]} = \mathbb{E}_u[X/n \mid X \geq \gamma_t],$$

where γ_t is the $(1-\rho)$-quantile of the $\mathsf{Bin}(n, v_{t-1})$ distribution.

Assume for concreteness that $n = 100$, $v = 1/6$, and $\gamma = 50$, which corresponds to $\ell \approx 1.702 \cdot 10^{-14}$. Table 2.1 presents the evolution of $\widehat{\gamma}_t$ and $\widehat{v}_t$ for $\rho = 0.05$, using sample size $N = 1000$. Note that iteration $t = 0$ corresponds to the original binomial pdf with success probability $u = 1/6$, while iterations $t = 1, \ldots, 5$ correspond to binomial pdfs with success probabilities $\widehat{v}_t$, $t = 1, \ldots, 5$, respectively.

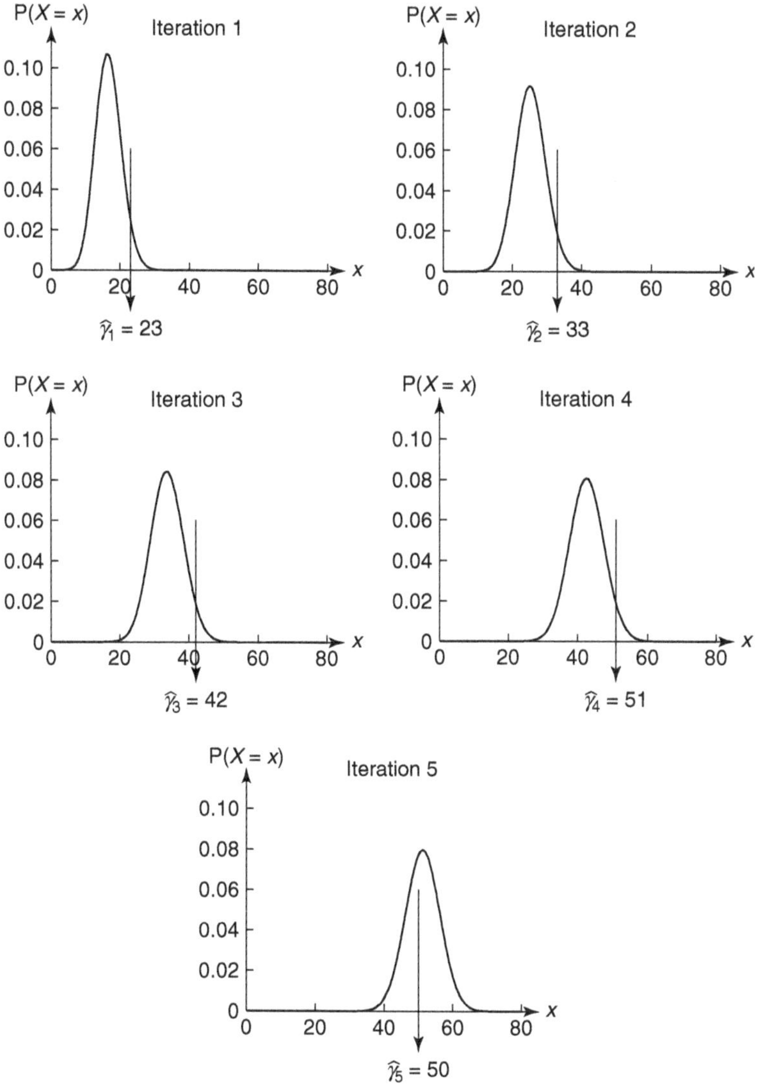

Figure 2.2 The five iterations of Algorithm 2.1. Iteration t samples from $\widehat{v}_{t-1}$, estimates level $\widehat{\gamma}_t$, and estimates the next sampling parameter $\widehat{v}_t$.

The final step of Algorithm 2.1 now invokes the importance sampling estimator (2.4) to estimate ℓ, using a sample size N_1 that is typically larger than N. Clearly, we reset the $\widehat{\gamma}_t$ obtained in the last iteration to γ.

Figure 2.2 illustrates the iterative procedure. We see that Algorithm 2.1 requires five iterations to reach the final level $\widehat{\gamma}_5 = 50$. Note that until iteration 4 both parameters $\widehat{\gamma}_t$ and $\widehat{v}_t$ increase gradually, each time "blowing up" the tail of the binomial pdf. At iteration five we reset the value $\widehat{\gamma}_4 = 51$ to the desired $\gamma = 50$ and estimate the optimal v value. □

Table 2.2 The evolution of γ_t and $\widehat{v}_t$ for the $\mathsf{Bin}(n = 100, u = 1/6)$ Example using $\gamma = 100$, $\rho = 0.01$ and $N = 1000$ samples

t	γ_t	$\widehat{v}_t$	t	γ_t	$\widehat{v}_t$
0	—	0.1667	5	73	0.7306
1	26	0.2710	6	82	0.8202
2	37	0.3739	7	91	0.9101
3	48	0.4869	8	97	0.9700
4	61	0.6108	9	100	1.0000

EXAMPLE 2.3 ***Degeneracy of v***

When γ is the maximum of $S(\boldsymbol{x})$, no "overshooting" of γ in Algorithm 2.1 will occur, and therefore γ_t does not need to be reset. In such cases, the sampling pdf may degenerate towards the atomic pdf that has all its mass concentrated at the points $\boldsymbol{x}$ where $S(\boldsymbol{x})$ is maximal. As an example, let $\gamma = 100$ be the target level in the above binomial example. In this case, clearly the final $\widehat{v}_t = 1$, which corresponds to the degenerate value of v. Table 2.2 and Figure 2.3 show the evolution of the parameters in the CE algorithm, using $\rho = 0.01$ and $N = 1000$. □

EXAMPLE 2.4

Consider again the stochastic shortest path graph of Figure 2.1. Let us take

$$\boldsymbol{u} = (1.0,\ 1.0,\ 0.3,\ 0.2,\ 0.1)$$

as the nominal parameter and estimate the probability ℓ that the minimum path length is greater than $\gamma = 6$.

Crude Monte Carlo with 10^8 samples—very large simulation effort—gave an estimate $8.01 \cdot 10^{-6}$ with an estimated relative error of 0.035.

To apply Algorithm 2.1 to this problem, we need to establish the updating rule for the reference parameter $\boldsymbol{v} = (v_1, \ldots, v_5)$. Since the components $X_1, \ldots, X_5$ are independent and form an exponential family parameterized by the mean we have [107]

$$\widehat{v}_{t,i} = \frac{\sum_{k=1}^{N} I_{\{S(\boldsymbol{X}_k) \geq \widehat{\gamma}_t\}} W(\boldsymbol{X}_k; \boldsymbol{u}, \widehat{\boldsymbol{v}}_{t-1})\, X_{ki}}{\sum_{k=1}^{N} I_{\{S(\boldsymbol{X}_k) \geq \widehat{\gamma}_t\}} W(\boldsymbol{X}_k; \boldsymbol{u}, \widehat{\boldsymbol{v}}_{t-1})}, \quad i = 1, \ldots, 5, \tag{2.15}$$

with $W(\boldsymbol{X}; \boldsymbol{u}, \boldsymbol{v})$ given in (2.10).

We set in all our experiments with Algorithm 2.1 the rarity parameter $\rho = 0.1$, the sample size for step 2–4 of the algorithm $N = 10^3$, and for the final sample size $N_1 = 10^5$. Table 2.3 displays the results of steps 1–5 of the CE algorithm.

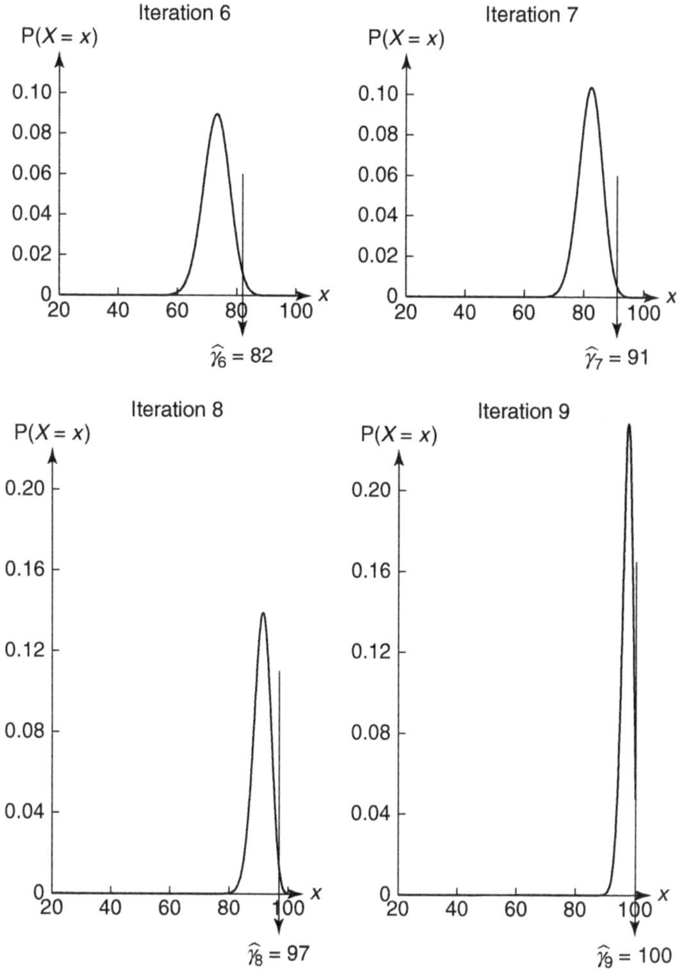

Figure 2.3 Degeneracy of the sampling distribution in Example 2.3.

We see that after five iterations, level $\gamma_5 = 6.0123$ is reached. As before, we set γ_5 to the desired $\gamma = 6.000$ and estimate the optimal parameter vector $\boldsymbol{v}$, which is $\widehat{\boldsymbol{v}}_6 = (6.7950, 6.7094, 0.2882, 0.1512, 0.1360)$. Using $\widehat{\boldsymbol{v}}_6$ we obtain $\widehat{\ell} = 7.85 \cdot 10^{-6}$, with an estimated RE of 0.035. The same RE was obtained for the crude Monte Carlo with 10^8 samples. Note that, whereas the standard method required more than an hour of computation time, the CE algorithm was finished in only 1 second, using a Matlab implementation on a 1500 MHz computer. We see that with a minimal amount of work we have achieved a dramatic reduction of the simulation effort. □

Table 2.3 Convergence of the sequence $\{(\widehat{\gamma}_t, \widehat{\boldsymbol{v}}_t)\}$ in Example 2.4.

t	$\widehat{\gamma}_t$			$\widehat{\boldsymbol{v}}_t$		
0		1.0000	1.0000	0.3000	0.2000	0.1000
1	1.1656	1.9805	2.0078	0.3256	0.2487	0.1249
2	2.1545	2.8575	3.0006	0.2554	0.2122	0.0908
3	3.1116	3.7813	4.0858	0.3017	0.1963	0.0764
4	4.6290	5.2803	5.6542	0.2510	0.1951	0.0588
5	6.0123	6.8001	6.7114	0.2913	0.1473	0.1384
6	6.0000	6.7950	6.7094	0.2882	0.1512	0.1360

2.3 CROSS-ENTROPY METHOD FOR OPTIMIZATION

Consider the following optimization problem:

$$\max_{\boldsymbol{x} \in \mathcal{X}} S(\boldsymbol{x}), \tag{2.16}$$

having $\mathcal{X}^*$ as the set of optimal solutions.

The problem is called a discrete or continuous optimization problem, depending on whether $\mathcal{X}$ is discrete or continuous. An optimization problem where some $\boldsymbol{x} \in \mathcal{X}$ are discrete and some others are continuous variables is called a mixed optimization problem. An optimization problem with a finite state space ($|\mathcal{X}| < \infty$) is called a combinatorial optimization problem. The latter will be the main focus of this section.

We will show here how the cross-entropy method can be applied for approximating a pdf f^* that is concentrated on $\mathcal{X}^*$. In fact, the iterations of the cross-entropy algorithm generate a sequence of pdfs f_t associated with a sequence of decreasing sets converging to $\mathcal{X}^*$.

Suppose that we set the optimal value $\gamma^* = \max_{\boldsymbol{x} \in \mathcal{X}} S(\boldsymbol{x})$, then an optimal pdf f^* is characterized by

$$\mathbb{P}_{f^*}(S(\boldsymbol{X}) \geq \gamma^*) = 1.$$

This matches exactly to the case of degeneracy of the cross-entropy algorithm for rare-event simulation in Example 2.3.

Hence, the starting point is to associate with the optimization problem (2.16) a meaningful estimation problem. To this end, we define a collection of indicator functions $\{I_{\{S(\boldsymbol{x}) \geq \gamma\}}\}$ on $\mathcal{X}$ for various levels $\gamma \in \mathbb{R}$. Next, let $\mathscr{F} = \{f(\cdot; \boldsymbol{v}), \boldsymbol{v} \in \mathcal{V}\}$ be a family of probability densities on $\mathcal{X}$, parameterized by real-valued parameters $\boldsymbol{v} \in \mathcal{V}$. For a fixed $\boldsymbol{u} \in \mathcal{V}$ we associate with (2.16) the problem of estimating the rare-event probability

$$\ell(\gamma) = \mathbb{P}_{\boldsymbol{u}}(S(\boldsymbol{X}) \geq \gamma^*) = \mathbb{E}_{\boldsymbol{u}}[I_{\{S(\boldsymbol{X}) \geq \gamma^*\}}]. \tag{2.17}$$

We call the estimation problem (2.17) the associated stochastic problem.

It is important to understand that one of the main goals of CE in optimization is to generate a sequence of pdfs $f(\cdot;\widehat{\boldsymbol{v}}_0), f(\cdot;\widehat{\boldsymbol{v}}_1), \ldots$, converging to a pdf $\widehat{f}$ that is close or equal to an optimal pdf f^*. For this purpose we apply Algorithm 2.1 for rare-event estimation, but without fixing in advance γ. It is plausible that if $\widehat{\gamma}^*$ is close to γ^*, then $\widehat{f}(\cdot) = f(\cdot;\widehat{\boldsymbol{v}}_T)$ assigns most of its probability mass on $\mathcal{X}^*$, and thus any $\boldsymbol{X}$ drawn from this distribution can be used as an approximation to an optimal solution $\boldsymbol{x}^*$, and the corresponding function value as an approximation to the true optimal γ^* in (2.16).

In summary, to solve a combinatorial optimization problem we will employ the CE Algorithm 2.1 for rare-event estimation without fixing γ in advance. By doing so, the CE algorithm for optimization can be viewed as a modified version of Algorithm 2.1. In particular, in analogy to Algorithm 2.1, we choose a not very small number ρ, say $\rho = 10^{-2}$, initialize the parameter $\boldsymbol{u}$ by setting $\boldsymbol{v}_0 = \boldsymbol{u}$, and proceed as follows:

1. **Adaptive updating of γ_t.** For a fixed $\boldsymbol{v}_{t-1}$, let γ_t be the $(1-\rho)$-quantile of $S(\boldsymbol{X})$ under $\boldsymbol{v}_{t-1}$. As before, an estimator $\widehat{\gamma}_t$ of γ_t can be obtained by drawing a random sample $\boldsymbol{X}_1, \ldots, \boldsymbol{X}_N$ from $f(\cdot;\boldsymbol{v}_{t-1})$ and then evaluating the sample $(1-\rho)$-quantile of the performances as
$$\widehat{\gamma}_t = S_{(\lceil (1-\rho)N \rceil)}. \tag{2.18}$$
2. **Adaptive updating of $\boldsymbol{v}_t$.** For fixed γ_t and $\boldsymbol{v}_{t-1}$, derive $\boldsymbol{v}_t$ from the solution of the program
$$\max_{\boldsymbol{v}} D(\boldsymbol{v}) = \max_{\boldsymbol{v}} \mathbb{E}_{\boldsymbol{v}_{t-1}}[I_{\{S(\boldsymbol{X})\geq\gamma_t\}} \log f(\boldsymbol{X};\boldsymbol{v})]. \tag{2.19}$$
The stochastic counterpart of (2.19) is as follows: for fixed $\widehat{\gamma}_t$ and $\widehat{\boldsymbol{v}}_{t-1}$, derive $\widehat{\boldsymbol{v}}_t$ from the following program:
$$\max_{\boldsymbol{v}} \widehat{D}(\boldsymbol{v}) = \max_{\boldsymbol{v}} \frac{1}{N}\sum_{k=1}^{N} I_{\{S(\boldsymbol{X}_k)\geq\widehat{\gamma}_t\}} \log f(\boldsymbol{X}_k;\boldsymbol{v}). \tag{2.20}$$

Note that, in contrast to the formulas (2.11) and (2.12) (for the rare-event setting), formulas (2.19) and (2.20) do not contain the likelihood ratio factors W. The reason is that in the rare-event setting, the initial (nominal) parameter $\boldsymbol{u}$ is specified in advance and is an essential part of the estimation problem. In contrast, the initial reference vector $\boldsymbol{u}$ in the associated stochastic problem is quite arbitrary. In effect, by dropping the likelihood ratio factor, we can efficiently estimate at each iteration t the CE optimal high-dimensional reference parameter vector $\boldsymbol{v}_t$ for the rare event probability $\mathbb{P}_{\boldsymbol{v}_t}(S(\boldsymbol{X}) \geq \gamma_t) \geq \mathbb{P}_{\boldsymbol{v}_{t-1}}(S(\boldsymbol{X}) \geq \gamma_t)$ by avoiding the degeneracy properties of the likelihood ratio.

Thus, the main CE optimization algorithm, which includes smoothed updating of parameter vector $\boldsymbol{v}$ and which presents a slight modification of Algorithm 2.1 can be summarized as follows.

Algorithm 2.2 ***Main CE Algorithm for Optimization***

1. ***Initialize:*** *Choose an initial parameter vector* $\boldsymbol{v}_0 = \widehat{\boldsymbol{v}}_0$. *Set* $t = 1$ *(level counter).*
2. ***Draw:*** *Generate a sample* $\boldsymbol{X}_1, \ldots, \boldsymbol{X}_N$ *from the density* $f(\cdot; \boldsymbol{v}_{t-1})$.
3. ***Select:*** *Compute the sample* $(1 - \rho)$*-quantile* $\widehat{\gamma}_t$ *of the performances according to* (2.18).
4. ***Update:*** *Use the same sample* $\boldsymbol{X}_1, \ldots, \boldsymbol{X}_N$ *and solve the stochastic program (2.20). Denote the solution by* $\tilde{\boldsymbol{v}}_t$.
5. ***Smooth:*** *Apply* (2.13) *to smooth out the vector* $\tilde{\boldsymbol{v}}_t$.
6. ***Iterate:*** *If the stopping criterion is met, stop; otherwise set* $t = t + 1$, *and return to Step 2.*

Note that when $S(\boldsymbol{x})$ must be minimized instead of maximized, we simply change the inequalities "$\geq$" to "$\leq$" and take the ρ-quantile instead of the $(1 - \rho)$-quantile. Alternatively, one can just maximize $-S(\boldsymbol{x})$.

Remark 2.4 ***Stopping Criteria***

As a stopping criterion, one can use, for example, if for some $t \geq d$, say $d = 5$,

$$\widehat{\gamma}_t = \widehat{\gamma}_{t-1} = \cdots = \widehat{\gamma}_{t-d}, \tag{2.21}$$

then stop.

As an alternative estimate for γ^*, one can consider

$$\tilde{\gamma}_T = \max_{0 \leq s \leq T} \widehat{\gamma}_s. \tag{2.22}$$

Before one implements the algorithm, one must decide on specifying the initial vector $\widehat{\boldsymbol{v}}_0$, the sample size N, the stopping parameter d, and the rarity parameter ρ. However, the execution of the algorithm is "self-tuning."

Note that the estimation Step 7 of Algorithm 2.1 is omitted in Algorithm 2.2, because, in the optimization setting, we are not interested in estimating ℓ per se.

Remark 2.5 ***Class of parametric densities***

To run the algorithm one needs to propose a class of parametric sampling densities $\{f(\cdot; \boldsymbol{v}), \boldsymbol{v} \in \mathcal{V}\}$, the initial vector $\widehat{\boldsymbol{v}}_0$, the sample size N, the rarity parameter ρ, and a stopping criterion. Of these, the most challenging is the selection of an appropriate class of parametric sampling densities $\{f(\cdot; \boldsymbol{v}), \boldsymbol{v} \in \mathcal{V}\}$. Typically, there is not a unique parametric family and the selection is guided by the following competing objectives. First, the class $\{f(\cdot; \boldsymbol{v}), \boldsymbol{v} \in \mathcal{V}\}$ has to be flexible enough to include a reasonable parametric approximation to an optimal importance sampling density g^* for the estimation of the associated rare-event probability ℓ. Second, each density $f(\cdot; \boldsymbol{v})$ has to be simple enough to allow fast random variable generation. In many cases, these two competing objectives are reconciled by using a standard statistical model for $f(\cdot; \boldsymbol{v})$, such as the multivariate Bernoulli or Gaussian densities with independent components of the vector $\boldsymbol{X} \sim f(\cdot; \boldsymbol{v})$.

Algorithm 2.2 can, in principle, be applied to any discrete optimization problem. However, for each individual problem, two essential ingredients need to be supplied:

1. We need to specify how the samples are generated. In other words, we need to specify the family of densities $\{f(\cdot; \boldsymbol{v})\}$.
2. We need to update the parameter vector $\boldsymbol{v}$, based on cross-entropy minimization program (2.20), which is the same for all optimization problems.

In general, there are many ways to generate samples from $\mathcal{X}$, and it is not always immediately clear which way of generating the sample will yield better results or easier updating formulas.

Remark 2.6 ***Parameter Selection***

The choice for the sample size N and the rarity parameter ρ depends on the size of the problem and the number of parameters in the associated stochastic problem. Typical choices are $\rho = 0.1$ or $\rho = 0.01$ and $N = c\,K$, where K is the number of parameters that need to be estimated/updated and c is a constant between 1 and 10.

Below we present several applications of the CE method to combinatorial optimization problems, namely to the multidimensional 0/1 knapsack problem, the mastermind game, and to reinforcement learning in Markov decision problems. We demonstrate numerically the efficiency of the CE method and its fast convergence for several case studies. For additional applications of CE, see [107].

2.3.1 The Multidimensional 0/1 Knapsack Problem

Consider the multidimensional 0/1 knapsack problem (see Section A.1.2):

$$\begin{aligned} &\max \sum_{j=1}^{n} p_j x_j \\ &\text{s.t.} \sum_{j=1}^{n} w_{ij} x_j \le c_i, \quad \text{for all } i = 1, \ldots, m \\ &x_j \in \{0, 1\}, \quad \text{for all } j = 1, \ldots, n. \end{aligned} \tag{2.23}$$

In other words, the goal is to maximize the performance function $S(\boldsymbol{x}) = \sum_{j=1}^{n} p_j x_j$ over the set $\mathcal{X}$ given by the knapsack constraints of (2.23).

Note that each item j is either chosen, $x_j = 1$, or left out, $x_j = 0$. We model this by a Bernoulli random variable $X_j \in \{0, 1\}$, with parameter v_j. By setting these variables independent, we get the following parameterized set:

$$\mathscr{F} = \{f(\boldsymbol{x}; \boldsymbol{v}) = \prod_{j=1}^{n} v_j^{x_j}(1 - v_j)^{1-x_j} : \boldsymbol{v} = (v_1, \ldots, v_n) \in [0, 1]^n\}$$

of Bernoulli pdfs. The optimal solution $\boldsymbol{x}^*$ of the knapsack problem (2.23) is characterized by a degenerate pdf $f^* \in \mathcal{F}$ for which all $v_j^* \in \{0, 1\}$; that is,

$$x_j^* = 1 \ \Rightarrow \ v_j^* = 1; \quad x_j^* = 0 \ \Rightarrow \ v_j^* = 0, \quad \text{for all } j = 1, \ldots, n.$$

To proceed, consider the stochastic program 2.20. We have

$$\begin{aligned}
&\max_{\boldsymbol{v}} \frac{1}{N} \sum_{k=1}^{N} I_k \log f(\boldsymbol{X}_k; \boldsymbol{v}) \\
\Leftrightarrow &\max_{\boldsymbol{v}} \sum_{k=1}^{N} I_k \log \prod_{j=1}^{n} v_j^{X_{kj}}(1 - v_j)^{1-X_{kj}} \\
\Leftrightarrow &\max_{\boldsymbol{v}} \sum_{k=1}^{N} I_k \sum_{j=1}^{n} (X_{kj} \log v_j + (1 - X_{kj}) \log (1 - v_j)) \\
\Leftrightarrow &\max_{\boldsymbol{v}} \sum_{j=1}^{n} \sum_{k=1}^{N} I_k \left(X_{kj} \log v_j + (1 - X_{kj}) \log (1 - v_j)\right),
\end{aligned}$$

where, for convenience, we write $I_k = I_{\{S(\boldsymbol{X}_k) \geq \widehat{\gamma}_t\}}$. Applying the first-order optimality conditions we have

$$\begin{aligned}
&\frac{\partial}{\partial v_j} \sum_{j=1}^{n} \sum_{k=1}^{N} I_k \left(X_{kj} \log v_j + (1 - X_{kj}) \log (1 - v_j)\right) = 0 \\
\Leftrightarrow \ &\sum_{k=1}^{N} I_k \frac{X_{kj}}{v_j} - \sum_{k=1}^{N} I_k \frac{1 - X_{kj}}{1 - v_j} = 0 \\
\Leftrightarrow \ &v_j = \frac{\sum_{k=1}^{N} I_k X_k}{\sum_{k=1}^{N} I_k} = \frac{\sum_{k=1}^{N} I_{\{S(\boldsymbol{X}_k) \geq \widehat{\gamma}_t\}} X_k}{\sum_{k=1}^{N} I_{\{S(\boldsymbol{X}_k) \geq \widehat{\gamma}_t\}}}
\end{aligned}$$

for $j = 1, \ldots, n$.

As a concrete example, consider the knapsack problem; available at http://people.brunel.ac.uk/~mastjjb/jeb/orlib/mknapinfo.html from the OR Library involving $n = 10$ variables and $m = 10$ constraints. In this problem the vector $\boldsymbol{p}$ is

$$\boldsymbol{p} = (600.1,\ 310.5,\ 1800.0,\ 3850.0,\ 18.6,\ 198.7,\ 882.0,\ 4200.0,\ 402.5,\ 327.0),$$

Table 2.4 The evolution of $\widehat{\gamma}_t$ and $\widehat{\boldsymbol{v}}_t$ for the 10 × 10-Knapsack Example, using $\rho = 0.05$, $\alpha = 0.8$, and $N = 1000$ samples

t	$\widehat{\gamma}_t$					$\widehat{v}_t$					
0	—	0.500	0.500	0.500	0.500	0.500	0.500	0.500	0.500	0.500	0.500
1	7528.3	0.580	0.628	0.644	0.356	0.516	0.596	0.452	0.900	0.292	0.596
2	8318.3	0.516	0.798	0.513	0.487	0.551	0.727	0.474	0.980	0.058	0.663
3	8377.0	0.119	0.768	0.103	0.897	0.638	0.609	0.095	0.996	0.012	0.517
4	8706.1	0.024	0.954	0.021	0.979	0.928	0.122	0.019	0.999	0.002	0.903
5	8706.1	0.005	0.991	0.004	0.996	0.986	0.024	0.004	1.000	0.000	0.981
7	8706.1	0.001	0.998	0.001	0.999	0.997	0.005	0.001	1.000	0.000	0.996
8	8706.1	0.000	1.000	0.000	1.000	0.999	0.001	0.000	1.000	0.000	0.999
9	8706.1	0.000	1.000	0.000	1.000	1.000	0.000	0.000	1.000	0.000	1.000

and the matrix $\{w_{ij}\}$ and the vector $\boldsymbol{c}$ are

$$\{w_{ij}\} = \begin{pmatrix} 20 & 5 & 100 & 200 & 2 & 4 & 60 & 150 & 80 & 40 \\ 20 & 7 & 130 & 280 & 2 & 8 & 110 & 210 & 100 & 40 \\ 60 & 3 & 50 & 100 & 4 & 2 & 20 & 40 & 6 & 12 \\ 60 & 8 & 70 & 200 & 4 & 6 & 40 & 70 & 16 & 20 \\ 60 & 13 & 70 & 250 & 4 & 10 & 60 & 90 & 20 & 24 \\ 60 & 13 & 70 & 280 & 4 & 10 & 70 & 105 & 22 & 28 \\ 5 & 2 & 20 & 100 & 2 & 5 & 10 & 60 & 0 & 0 \\ 45 & 14 & 80 & 180 & 6 & 10 & 40 & 100 & 20 & 0 \\ 55 & 14 & 80 & 200 & 6 & 10 & 50 & 140 & 30 & 40 \\ 65 & 14 & 80 & 220 & 6 & 10 & 50 & 180 & 30 & 50 \end{pmatrix} \quad \boldsymbol{c} = \begin{pmatrix} 450 \\ 540 \\ 200 \\ 360 \\ 440 \\ 480 \\ 200 \\ 360 \\ 440 \\ 480, \end{pmatrix}$$

The performance of the CE algorithm is illustrated in Table 2.4 using the sample size $N = 1000$, rarity parameter $\rho = 0.05$, and smoothing parameter $\alpha = 0.8$. The initial probabilities were all set to $v_j = 0.5$. Observe that the probability vector $\widehat{\boldsymbol{v}}_t$ quickly converges to a degenerate one, corresponding to the solution $\boldsymbol{x} = (0, 1, 0, 1, 1, 0, 0, 1, 0, 1)$ with value $S(\boldsymbol{x}) = 8706.1$, which is also the optimal known value $\gamma^* = 8706.1$. Hence, the optimal solution has been found.

2.3.2 Mastermind Game

In the well-known game Mastermind, the objective is to decipher a hidden "code" of colored pegs through a series of guesses. After each guess, new information on the true code is provided in the form of black and white pegs. A black peg is earned for each peg that is exactly right and a white peg for each peg that is in the solution, but in the wrong position; see Figure 2.4. To get a feel for the game, one can visit www.archimedes-lab.org/mastermind.html and play the game.

Consider a Mastermind game with m colors, numbered $\{1, \ldots, m\}$ and n pegs (positions). The hidden solution and a guess can be represented by a row vector of length n with numbers in $\{1, \ldots, m\}$. For example, for $n = 5$ and $m = 7$ the

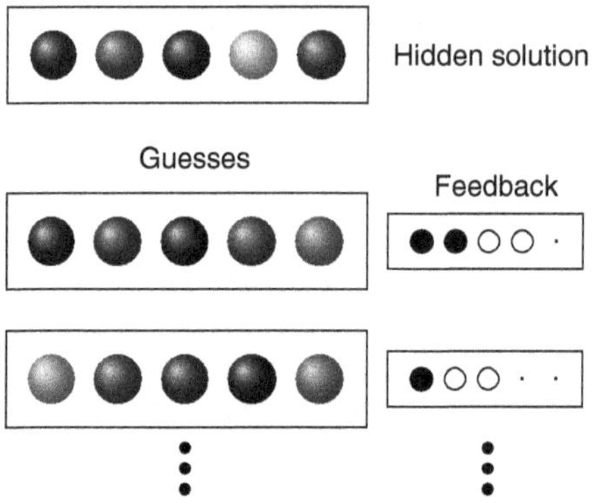

Figure 2.4 The Mastermind game.

solution could be $\boldsymbol{y} = (4, 2, 4, 7, 3)$, and a possible guess $\boldsymbol{x} = (4, 3, 4, 2, 1)$. Let $\mathcal{X}$ be the space of all possible guesses; clearly $|\mathcal{X}| = m^n$. On $\mathcal{X}$ we define a performance function S that returns for each guess $\boldsymbol{x}$ the "pegscore"

$$S(\boldsymbol{x}) = 2 * N_{\text{BlackPegs}}(\boldsymbol{x}) + N_{\text{WhitePegs}}(\boldsymbol{x}), \tag{2.24}$$

where $N_{\text{BlackPegs}}(\boldsymbol{x})$ and $N_{\text{WhitePegs}}(\boldsymbol{x})$ are the number of black and white pegs returned after the guess, $\boldsymbol{x}$. We have assumed, somewhat arbitrarily, that a black peg is worth twice as much as a white peg. As an example, for the solution $\boldsymbol{y} = (4, 2, 4, 7, 3)$ and guess $\boldsymbol{x} = (4, 3, 4, 2, 1)$ above one gets two black pegs (for the first and third pegs in the guess) and two white pegs (for the second and fourth pegs in the guess, which match the fifth and second pegs of the solution but are in the wrong place). Hence the score for this guess is 6.

There are some specially designed algorithms that efficiently solve the problem for different number of colors and pegs. For examples, see www.mathworks.com/contest/mastermind.cgi/home.html.

Note that, with the above performance function, the problem can be formulated as an optimization problem, that is, as $\max_{\boldsymbol{x}} S(\boldsymbol{x})$, and hence we could apply the CE method to solve it. In order to apply the CE method we first need to generate random guesses $\boldsymbol{X} \in \mathcal{X}$ using an $n \times m$ probability matrix P. Each element p_{ij} of P describes the probability that we choose the jth color for the ith peg (location). Since only one color may be assigned to one peg, we have that

$$\sum_{j=1}^{m} p_{ij} = 1, \quad \forall i. \tag{2.25}$$

In each stage t of the CE algorithm, we sample independently for each row (that is, each peg) a color using a probability matrix $P = \widehat{P}_t$ and calculate the score S according to (2.24). The updating of the elements $\widehat{p}_{t,ij}$ of the probability matrix

Table 2.5 Dynamics of Algorithm 2.2 for the mastermind problem with $m = 33$ colors and $n = 36$ pegs. In each iteration S_t was the sampled maximal score

t	S_t	$\widehat{\gamma}_t$
1	31	26
2	34	28
3	38	32
4	42	35
5	48	40
6	55	45
7	60	51
8	64	57
9	70	63
10	72	68
11	72	72

$\widehat{P}_t$ is performed according to

$$p_{t,ij} = \frac{\sum_{k=1}^{N} I_{\{S(\boldsymbol{X}_k) \geq \gamma_t\}} I_{\{X_{ki}=j\}}}{\sum_{k=1}^{N} I_{\{S(\boldsymbol{X}_k) \geq \widehat{\gamma}_t\}}}, \tag{2.26}$$

where the samples $\boldsymbol{X}_1, \ldots, \boldsymbol{X}_N$ are drawn independently and identically distributed using matrix $\widehat{P}_{t-1}$. Note that in this case $\widehat{p}_{t,ij}$ in (2.26) simply counts the fraction of times among the elites that color j has been assigned to peg i.

2.3.2.1 Simulation Results

Table 2.5 represents the dynamics of the CE algorithm for the Mastermind test problem denoted at www.mathworks.com/contest/mastermind.cgi/home.html as "problem (5.2.3)" with the matrix P of size $n \times m = 36 \times 33$. We used $N = 5\,m\,n = 5940$, $\rho = 0.01$, and $\alpha = 0.7$. It took 34 seconds of CPU time to find the true solution.

2.3.3 Markov Decision Process and Reinforcement Learning

The Markov decision process (MDP) model is standard in machine learning, operations research, and related fields. The application to reinforcement learning in this section is based on [83]. We start by reviewing briefly some of the basic definitions and concepts in MDP. For details see, for example, [95].

An MDP is defined by a tuple $(\mathcal{Z}, \mathcal{A}, P, r)$ where

1. $\mathcal{Z} = \{1, \ldots, n\}$ is a finite set of states.

2. $\mathcal{A} = \{1, \ldots, m\}$ is the set of possible actions by the decision maker. We assume it is the same for every state, to ease notations.
3. P is the transition probability matrix with elements $P(z'|z, a)$ presenting the transition probability from state z to state z', when action a is chosen.
4. $r(z, a)$ is the immediate reward for performing action a in state z (r may be a random variable).

At each time instance k, the decision maker observes the current state z_k and determines the action to be taken (say a_k). As a result, a reward given by $r(z_k, a_k)$ is received immediately, and a new state z' is chosen according to $P(z'|z_k, a_k)$. A policy or strategy π is a rule that determines, for each history $H_k = z_0, a_0, \ldots, z_{k-1}, a_{k-1}, z_k$ of states and actions, the probability distribution of the decision maker's actions at time k. A policy is called Markov if each action depends only on the current state z_k. A Markov policy is called stationary if it does not depend on the time k. If the decision rule π of a stationary policy is nonrandomized, we say that the policy is deterministic.

In the sequel, we consider problems for which an optimal policy belongs to the class of stationary Markov policies. Associated with such a MDP are the stochastic processes of consecutive observed states $Z_0, Z_1, \ldots$ and actions $A_0, A_1, \ldots$. For stationary policies we may denote $A_k = \pi(Z_k)$, which is a random variable on the action set $\mathcal{A}$. In case of deterministic policies it degenerates in a single action. Finally, we denote the immediate rewards in state Z_k by $R_k = r(Z_k, A_k) = R_k(Z_k, \pi(Z_k))$.

The goal of the decision maker is to maximize a certain reward function. The following are standard reward criteria:

1. *Finite horizon reward.* This applies when there exists a finite time τ (random or deterministic) at which the process terminates. The objective is to maximize the total reward
$$S(\pi) = \mathbb{E}_\pi \sum_{k=0}^{\tau-1} R_k \,. \tag{2.27}$$
Here, $\mathbb{E}_\pi$ denotes the expectation with respect to the probability measure induced by the strategy π.
2. *Infinite horizon discounted reward.* The objective is to find a strategy π that maximizes
$$S(\pi) = \mathbb{E}_\pi \sum_{k=0}^{\infty} \beta^k \, R_k, \tag{2.28}$$
where $0 < \beta < 1$ is the discount factor.
3. *Average reward.* The objective is to maximize
$$S(\pi) = \liminf_{\tau\to\infty} \frac{1}{\tau} \, \mathbb{E}_\pi \sum_{k=0}^{\tau-1} R_k.$$

Suppose that the MDP is identified; that is, the transition probabilities and the immediate rewards are specified and known. Then it is well known that there exists a deterministic stationary optimal strategy for the above three cases. Moreover, there are several efficient methods for finding it, such as value iteration and policy iteration [95].

In the next section, we will restrict our attention to the stochastic shortest path MDP, where it is assumed that the process starts from a specific initial state $z_0 = z^{\text{start}}$ and terminates in an absorbing state z^{fin} with zero reward. For the finite horizon reward criterion τ is the time at which z^{fin} is reached (which we will assume will always happen eventually). It is well known that for the shortest path MDP there exists a stationary Markov policy that maximizes $S(\pi)$ for each of the reward functions above.

However, when the transition probability or the reward in an MDP are unknown, the problem is much more difficult and is referred to as a learning probability. A well-known framework for learning algorithms is reinforcement learning (RL), where an agent learns the behavior of the system through trial-and-error with an unknown dynamic environment; see [61].

There are several approaches to RL, which can be roughly divided into the following three classes: model-based, model-free, and policy search. In the model-based approach, a model of the environment is first constructed. The estimated MDP is then solved using standard tools of dynamic programming (see [66]).

In the model-free approach, one learns a utility function instead of learning the model. The optimal policy is to choose at each state an action that maximizes the expected utility. The popular Q-learning algorithm (e.g., [125]) is an example of this approach.

In the policy search approach, a subspace of the policy space is searched, and the performance of policies is evaluated based on their empirical performance (e.g., [7]). An example of a gradient-based policy search method is the REINFORCE algorithm of [127]. For a direct search in the policy space approach see [98]. The CE algorithm in the next section belongs to the policy search approach.

Remark 2.7 ***Stochastic Approximation***

Many RL algorithms are based on the classic stochastic approximation (SA) algorithm. To explain SA, assume that we need to find the unique solution v^* of a nonlinear equation $\mathbb{E}_v[S] = 0$, where S is a random variable with a pdf parameterized by v, such that its expectation $\ell(v) = \mathbb{E}_v[S]$ is not available in closed form. However, assume that we can simulate S for any parameter v. The SA algorithm for estimating v^* is

$$v_{k+1} = v_k + \beta_k S(v_k),$$

where β_k is a positive sequence satisfying

$$\sum_{k=1}^{\infty} \beta_k = \infty, \quad \sum_{k=1}^{\infty} \beta_k^2 < \infty. \tag{2.29}$$

The connection between SA and Q-learning is given by [117]. This work has made an important impact on the entire field of RL. Even if β_k remains bounded away from 0 (and thus convergence is not guaranteed), it is still required that β_k is small in order to ensure convergence to a reasonable neighboring solution [10].

Our main goal here is to introduce a fast learning algorithm based on the CE method instead of the slow SA algorithms and to demonstrate its high efficiency.

2.3.3.1 Policy Learning via the CE Method

We consider a CE learning algorithm for the shortest path MDP, where it is assumed that the process starts from a specific initial state z_0, and that there is an absorbing state z^{fin} with zero reward. The objective is given in (2.27), with τ being the stopping time at which z^{fin} is reached (which we will assume will always happen).

To put this problem in the CE framework, consider the maximization problem (2.27). Recall that for the shortest path MDP an optimal stationary strategy exists. We can represent each stationary strategy as a vector $\boldsymbol{x} = (x_1, \dots, x_n)$ with $x_i \in \{1, \dots, m\}$ being the action taken when visiting state i. Writing the expectation in (2.27) as

$$S(\boldsymbol{x}) = \mathbb{E}_{\boldsymbol{x}} \sum_{k=0}^{\tau-1} r(Z_k, A_k), \tag{2.30}$$

we see that the optimization problem (2.27) is of the form $\max_{\boldsymbol{x}\in\mathcal{X}} S(\boldsymbol{x})$. We shall also consider the case where $S(\boldsymbol{x})$ is measured (observed) with some noise, in which case we have a noisy optimization problem.

The idea now is to combine the random policy generation and the random trajectory generation in the following way: At each stage of the CE algorithm we generate random policies and random trajectories using an auxiliary $n \times m$ matrix $P = (p_{za})$, such that for each state z we choose action a with probability p_{za}. Once this "policy matrix" P is defined, each iteration of the CE algorithm comprises the following two standard phases:

1. Generation of N random trajectories $(Z_0, A_0, Z_1, A_1, \dots, Z_\tau, A_\tau)$ using the auxiliary policy matrix P. The cost of each trajectory is computed via

$$\widehat{S}(\boldsymbol{X}) = \sum_{j=0}^{\tau-1} r(Z_k, A_k). \tag{2.31}$$

2. Updating of the parameters of the policy matrix (p_{za}) on the basis of the data collected in the first phase.

The matrix P is typically initialized to a uniform matrix ($p_{ij} = 1/m$.) We describe both the trajectory generation and updating procedure in more detail. We shall show that in calculating the associated sample performance, one can take into account the Markovian nature of the problem and speed up the Monte Carlo process.

2.3.3.2 Generating Random Trajectories

Generation of random trajectories for MDP is straightforward and is given for convenience only. All one has to do is to start the trajectory from the initial state $z_0 = z^{\text{start}}$ and follow the trajectory by generating at each new state according to the probability distribution of P, until the absorbing state z^{fin} is reached at, say, time τ,

Algorithm 2.3 ***Trajectory Generation for MDP***

Input: P auxiliary policy matrix.

1. *Start from the given initial state $Z_0 = z^{start}$, set $k = 0$.*
2. *Generate an action A_k according to the Z_kth row of P, calculate the reward $r_k = r(Z_k, A_k)$, and generate a new state Z_{k+1} according to $P(\bullet \mid Z_k, A_k)$. Set $k = k + 1$. Repeat until $z_k = z^{fin}$.*
3. *Output the total reward of the trajectory $(Z_0, A_0, Z_1, A_1, \ldots, Z_\tau)$ given by* (2.31).

2.3.3.3 Updating Rules

Let the N trajectories $\boldsymbol{X}_1, \ldots, \boldsymbol{X}_N$, their scores, $S(\boldsymbol{X}_1), \ldots, S(\boldsymbol{X}_N)$, and the associated $(1-\rho)$-quantile $\widehat{\gamma}_t$ of their order statistics be given. Then one can update the parameter matrix (p_{za}) using the CE method, namely as per

$$p_{t,za} = \frac{\sum_{k=1}^{N} I_{\{S(\boldsymbol{X}_k)\geq\gamma_t\}}\, I_{\{\boldsymbol{X}_k\in\mathcal{X}_{za}\}}}{\sum_{k=1}^{N} I_{\{S(\boldsymbol{X}_k)\geq\widehat{\gamma}_t\}} I_{\{\boldsymbol{X}_k\in\mathcal{X}_z\}}}, \tag{2.32}$$

where the event $\{\boldsymbol{X}_k \in \mathcal{X}_z\}$ means that the trajectory $\boldsymbol{X}_k$ contains a visit to state z and the event $\{\boldsymbol{X}_k \in \mathcal{X}_{za}\}$ means the trajectory corresponding to policy $\boldsymbol{X}_k$ contains a visit to state z in which action a was taken.

We now explain how to take advantage of the Markovian nature of the problem. Let us think of a maze where a certain trajectory starts badly, that is, the path is not efficient in the beginning, but after some time it starts moving quickly toward, the goal. According to (2.32), all the updates are performed in a similar manner in every state in the trajectory. However, the actions taken in the states that were sampled near the target were successful, so one would like to "encourage" these actions. Using the Markovian property one can substantially improve the above algorithm by considering for each state the part of the reward from the visit to that state onward. We therefore use the same trajectory and simultaneously calculate the performance for every state in the trajectory separately. The idea is that each choice of action in a given state affects the reward from that point on, regardless of the past.

The sampling Algorithm 2.3 does not change in steps 1 and 2. The difference is in step 3. Given a policy $\boldsymbol{X}$ and trajectory $(Z_0, A_0, Z_1, A_1, \ldots, Z_\tau, A_\tau)$, we calculate the performance from every state until termination. For every state $z = Z_j$

in the trajectory, the (estimated) performance is $\widehat{S}_z(\boldsymbol{X}) = \sum_{k=j}^{\tau-1} r_k$. The updating formula for $\widehat{p}_{za}$ is similar to (2.32), however, each state z is updated separately according to the (estimated) performance $\widehat{S}_z(\boldsymbol{X})$ obtained from state z onward.

$$\widehat{p}_{t,za} = \frac{\sum_{k=1}^{N} I_{\{\widehat{S}_z(\boldsymbol{X}_k)\geq\widehat{\gamma}_{t,z}\}} I_{\{\boldsymbol{X}_k\in\mathcal{X}_{za}\}}}{\sum_{k=1}^{N} I_{\{\widehat{S}_z(\boldsymbol{X}_k)\geq\widehat{\gamma}_{t,z}\}} I_{\{\boldsymbol{X}_k\in\mathcal{X}_z\}}} . \tag{2.33}$$

It is important to understand here that, in contrast to (2.32), the CE optimization is carried for every state separately and a different threshold parameter $\widehat{\gamma}_{t,z}$ is used for every state z, at iteration t. This facilitates faster convergence for "easy" states where the optimal strategy is easy to find. The above trajectory sampling method can be viewed as a variance reduction method. Numerical results indicate that the CE algorithm with updating (2.33) is much faster than that with updating (2.32).

2.3.3.4 Numerical Results with the Maze Problem

The CE Algorithm with the updating rule (2.33) and trajectory generation according to Algorithm 2.3 was tested for a stochastic shortest path MDP and, in particular, for a maze problem, which presents a two-dimensional grid world. The agent moves on a grid in four possible directions. The goal of the agent is to move from the upper-left corner (the starting state) to the lower-right corner (the goal) in the shortest path possible. The maze contains obstacles (walls) into which movement is not allowed. These are modeled as forbidden moves for the agent.

We assume the following:

1. The moves in the grid are allowed in four possible directions with the goal to move from the upper-left corner to the lower-right corner.
2. The maze contains obstacles ("walls") into which movement is not allowed.
3. The cost of allowed moves are random variables uniformly distributed between 0.5 and 1.5. The cost of forbidden moves are random variables uniformly distributed between 25 and 75. Thus, the expected costs are 1 and 50 for the allowed and forbidden moves, respectively.

In addition, we introduce the following:

- A small (failure) probability not to succeed moving in an allowed direction
- A small probability of succeeding moving in the forbidden direction ("moving through the wall")

In Figure 2.5 we present the results for 20×20 maze. We set the following parameters: $N = 1000$, $\rho = 0.03$, $\alpha = 0.7$.

The initial policy was a uniformly random one. The success probabilities in the allowed and forbidden states were taken to be 0.95 and 0.05, respectively. The arrows $z \to z'$ in Figure 2.5 indicate that at the current iteration the probability of going from z to z' is at least 0.01. In other words, if a corresponds to the action

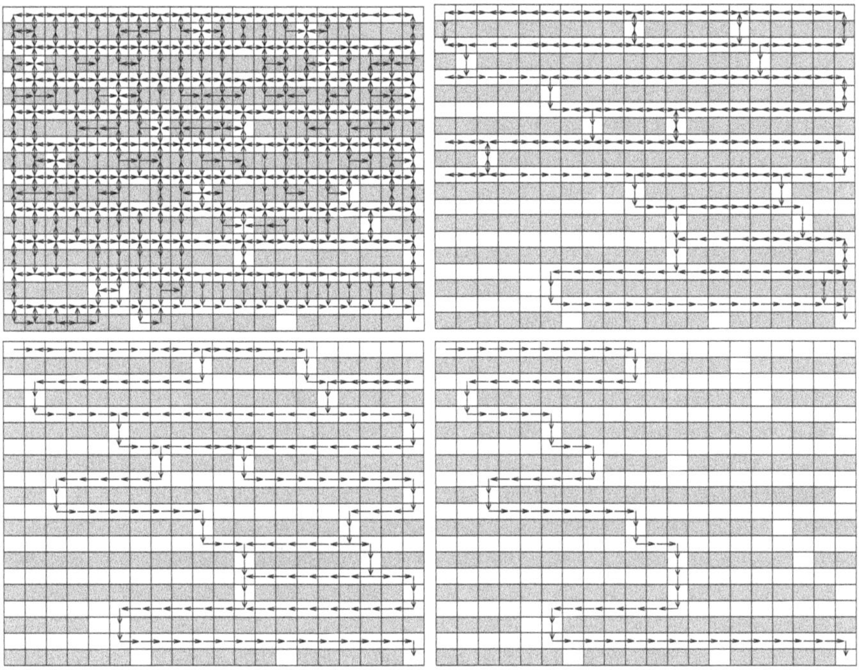

Figure 2.5 Performance of the CE Algorithm for the 20 × 20 maze. Each arrow indicates a probability > 0.01 of going in that direction.

that will lead to state z' from z then we will plot an arrow from z to z' provided $p_{za} > 0.01$.

In all our experiments, CE found the target exactly, within 5–10 iterations, and CPU time was less than one minute (on a 500MHz Pentium processor). Note that the successive iterations of the policy in Figure 2.5 quickly converge to the optimal policy. We have also run the algorithm for several other mazes and the optimal policy was always found.

2.4 CONTINUOUS OPTIMIZATION

When the state space is continuous, in particular, when $\mathcal{X} = \mathbb{R}^n$, the optimization problem (2.16) is often referred to as a continuous optimization problem. The CE sampling distribution on $\mathbb{R}^n$ can be quite arbitrary and does not need to be related to the function that is being optimized. The generation of a random vector $\boldsymbol{X} = (X_1, \ldots, X_n) \in \mathbb{R}^n$ in Step 2 of Algorithm 2.2 is most easily performed by drawing the n coordinates independently from some two-parameter distribution. In most applications a normal (Gaussian) distribution is employed for each component. Thus, the sampling density $f(\cdot; \boldsymbol{v})$ of $\boldsymbol{X}$ is characterized by a vector of means $\boldsymbol{\mu}$ and a vector of variances $\boldsymbol{\sigma}^2$ (and we may write $\boldsymbol{v} = (\boldsymbol{\mu}, \boldsymbol{\sigma}^2)$). The choice of the normal distribution is motivated by the availability of fast normal random number generators on modern statistical software and the fact that the cross-entropy minimization yields a very simple solution—at each iteration of the CE algorithm the parameter vectors $\boldsymbol{\mu}$ and $\boldsymbol{\sigma}^2$ are the vectors of sample means and sample variance of the elements of the set of N^{elite} best-performing vectors (that is, the elite set); see, for example, [71]. In summary, the CE method for continuous optimization with a Gaussian sampling density is as follows.

Algorithm 2.4 ***CE for Continuous Optimization: Normal Updating***

1. ***Initialize:*** *Choose* $\widehat{\boldsymbol{\mu}}_0$ *and* $\widehat{\boldsymbol{\sigma}}_0^2$. *Set* $t = 1$.
2. ***Draw:*** *Generate a random sample* $\mathbf{X}_1, \ldots, \mathbf{X}_N$ *from the* $\mathsf{N}(\widehat{\boldsymbol{\mu}}_{t-1}, \widehat{\boldsymbol{\sigma}}_{t-1}^2)$ *distribution.*
3. ***Select:*** *Let* $\mathcal{I}$ *be the indices of the* N^{elite} *best performing (elite) samples.*
4. ***Update:*** *For all* $j = 1, \ldots, n$ *let*

$$\tilde{\mu}_{t,j} = \sum_{k \in \mathcal{I}} X_{k,j} / N^{\text{elite}} \tag{2.34}$$

$$\tilde{\sigma}_{t,j}^2 = \sum_{k \in \mathcal{I}} (X_{k,j} - \tilde{\mu}_{t,j})^2 / N^{\text{elite}}. \tag{2.35}$$

5. ***Smooth:***

$$\widehat{\boldsymbol{\mu}}_t = \alpha \tilde{\boldsymbol{\mu}}_t + (1 - \alpha) \widehat{\boldsymbol{\mu}}_{t-1}, \quad \widehat{\boldsymbol{\sigma}}_t = \beta_t \tilde{\boldsymbol{\sigma}}_t + (1 - \beta_t) \widehat{\boldsymbol{\sigma}}_{t-1} . \tag{2.36}$$

6. ***Iterate:*** *Stop if* $\max_j \{\widehat{\sigma}_{t,j}\} < \varepsilon$, *and return* $\boldsymbol{\mu}_t$ *(or the overall best solution generated by the algorithm) as the approximate solution to the optimization. Otherwise, increase* t *by* 1 *and return to Step 2.*

Smoothing, as in Step 5, is often crucial to prevent premature shrinking of the sampling distribution. Instead of using a single smoothing factor, it is often useful to use separate smoothing factors for $\widehat{\boldsymbol{\mu}}_t$ and $\widehat{\boldsymbol{\sigma}}_t$ (see (2.36)). In particular, it is suggested in [108] to use the following dynamic smoothing for $\widehat{\boldsymbol{\sigma}}_t$:

$$\beta_t = \zeta - \zeta \left(1 - \frac{1}{t}\right)^q \tag{2.37}$$

where q is an integer (typically between 5 and 10) and ζ is a smoothing constant (typically between 0.8 and 0.99). Finally, significant speed-up, can be achieved by using a parallel implementation of CE [46].

Another approach is to inject extra variance into the sampling distribution, for example, by increasing the components of $\boldsymbol{\sigma}^2$, once the distribution has degenerated; see the examples below and [11].

EXAMPLE 2.5 ***The Peaks Function***

MATLAB's `peaks` function

$$S(x) = 3\,(1 - x_1)^2 \,\exp\,(-(x_1^2) - (x_2 + 1)^2)$$

$$- 10\ (x_1/5 - x_1^3 - x_2^5)\ \exp\ (-x_1^2 - x_2^2)$$
$$- 1/3\ \exp\ (-(x_1+1)^2 - x_2^2)$$

has various local maxima. In this case, implementing the CE Algorithm 2.4 we found that the global maximum of this function is approximately $\gamma^* = 8.10621359$ and is attained at $\boldsymbol{x}^* = (-0.0093151, 1.581363)$. The choice of the initial value for $\boldsymbol{\mu}$ is not important, but the initial standard deviations should be chosen large enough to ensure initially a "uniform" sampling of the region of interest. The CE algorithm is stopped when all standard deviations of the sampling distribution are less than some small ε.

Figure 2.6 gives the evolution of the worst and the best of the elite samples, that is, $\widehat{\gamma}_t$ and S_t^*, for each iteration t. We see that the values quickly converge to the optimal value γ^*. □

Remark 2.8 ***Injection***

When using the CE method to solve practical optimization problems with many constraints and many local optima, it is sometimes necessary to prevent the sampling distribution from shrinking too quickly. A simple but effective approach is the following injection method [107]. Let S_t^* denote the best performance found at the t-th iteration, and (in the normal case) σ_t^* the largest standard deviation at the t-th iteration. If σ_t^* is sufficiently small and $|S_t^* - S_{t-1}^*|$ is also small, then add some small value to each standard deviation, for example, a constant δ or the value $c\ |S_t^* - S_{t-1}^*|$, for some fixed δ and c. When using CE with injection, a possible stopping criterion is to stop when a certain number of injections is reached.

2.5 NOISY OPTIMIZATION

One of the distinguishing features of the CE Algorithm 2.2 is that it can easily handle noisy optimization problems, that is, when the objective function $S(\boldsymbol{x})$ is corrupted with noise. We denote such noisy function as $\widehat{S}(\boldsymbol{x})$ and assume that for each $\boldsymbol{x}$ we can readily obtain an outcome of $\widehat{S}(\boldsymbol{x})$, for example, via generation of some additional random vector $\boldsymbol{Y}$, whose distribution may depend on $\boldsymbol{x}$.

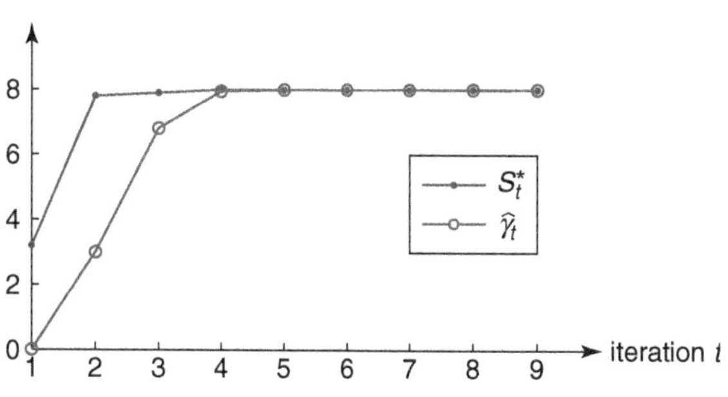

Figure 2.6 Evolution of the CE algorithm for the peaks function.

A classical example of noisy optimization is simulation-based optimization [107]. A typical instance is the buffer allocation problem (BAP). The objective in the BAP is to allocate n buffer spaces among the $m-1$ "niches" (storage areas) between m machines in a serial production line, so as to optimize some performance measure, such as the steady-state throughput. This performance measure is typically not available analytically and thus must be estimated via simulation. A detailed description of the BAP with application of CE is given in [107].

Another example is the noisy traveling salesman problem (TSP), where, say, the cost matrix (c_{ij}), denoted now by $\boldsymbol{Y} = (Y_{ij})$, is random. Think of Y_{ij} as the "random" time to travel from city i to city j.

The total cost of a tour $\boldsymbol{x} = (x_1, \ldots, x_n)$ is given by

$$\widehat{S}(\boldsymbol{x}) = \sum_{i=1}^{n-1} Y_{x_i,x_{i+1}} + Y_{x_n,x_1}. \tag{2.38}$$

We assume that $\mathbb{E}[Y_{ij}] = c_{ij}$.

The main CE optimization Algorithm 2.2 for deterministic functions $S(\boldsymbol{x})$ is valid also for noisy ones $\widehat{S}(\boldsymbol{x})$. Extensive numerical studies [107] with such noisy Algorithm 2.2 show that it works nicely because, during the course of optimization, it filters out efficiently the noise component from $\widehat{S}(\boldsymbol{x})$.

We applied our CE Algorithm 2.2 to both deterministic and noisy TSP. We selected a number of benchmark problems from the TSP library (www.iwr.uni-heidelberg.de/groups/comopt/software/TSPLIB95/tsp/). In all cases, the same set of CE parameters were chosen: $\rho = 0.03$, $\alpha = 0.7$, $N = 5 \cdot n^2$. Table 2.6 presents the performance of the deterministic version of Algorithm 2.2 for several TSPs from this library. To study the variability in the solutions, each problem was repeated 10 times. In Table 2.6, "min", "mean", and "max" denote the smallest (that is, best), average, and largest of the 10 estimates for the optimal value. The best-known solution is denoted by γ^*.

The average CPU time in seconds and the average number of iterations are given in the last two columns. The size of the problem (number of nodes) is indicated in its name. For example, st70 has $n = 70$ nodes.

EXAMPLE 2.6 ***Noisy TSP***

Suppose that in the first test case of Table 2.6, `burma14`, some uniform noise is added to the cost matrix. In particular, suppose that the cost of traveling from i to j is given by $Y_{ij} \sim \mathrm{U}(c_{ij} - 8, c_{ij} + 8)$, where c_{ij} is the cost for the deterministic case. The expected cost is thus $\mathbb{E}[Y_{ij}] = c_{ij}$, and the total cost $\widehat{S}(\boldsymbol{x})$ of a tour $\boldsymbol{x}$ is given by (2.38). The CE algorithm for optimizing the unknown $S(\boldsymbol{x}) = \mathbb{E}[\widehat{S}(\boldsymbol{x})]$ remains exactly the same as in the deterministic case, except that $S(\boldsymbol{x})$ is replaced with $\widehat{S}(\boldsymbol{x})$, and that a different stopping criterion than (2.21) needs to be employed. A simple rule is to stop when the transition probabilities $\widehat{p}_{t,ij}$ satisfy $\min(\widehat{p}_{t,ij}, 1 - \widehat{p}_{t,ij}) < \varepsilon$ for all i and j. An alternative stopping rule taken from [107] is as follows.

Table 2.6 Case studies for the TSP

file	γ^*	min	mean	max	CPU	$\bar{T}$
burma14	3323	3323	3325.6	3336	0.14	12.4
ulysses16	6859	6859	6864.0	6870	0.21	14.1
ulysses22	7013	7013	7028.9	7069	1.18	22.1
bayg29	1610	1610	1628.6	1648	4.00	28.2
bays29	2020	2020	2030.9	2045	3.83	27.1
dantzig42	699	706	717.7	736	19.25	38.4
eil51	426	428	433.9	437	65.00	63.4
berlin52	7542	7618	7794	8169	64.55	59.9
st70	675	716	744.1	765	267.50	83.7
eil76	538	540	543.5	547	467.30	109.0
pr76	108159	109882	112791.0	117017	375.30	88.9

2.5.1 Stopping Criterion

To identify the stopping time T, denote by $\widehat{\gamma}_{1t}$ and $\widehat{\gamma}_{2t}$ the worst of the elite samples $(\widehat{\gamma}_t)$ for both the deterministic and noisy case, respectively, and we consider the following moving average-process:

$$B_t(k) = \frac{1}{k} \sum_{s=t-k+1}^{t} \widehat{\gamma}_{2s}, \quad t = s, s+1, \ldots, \quad s \geq k, \tag{2.39}$$

where k is fixed, say $k = 10$. Define next

$$B_t^-(k, r) = \min_{j=1,\ldots,r} B_{t+j}(k) \tag{2.40}$$

and

$$B_t^+(k, r) = \max_{j=1,\ldots,r} B_{t+j}(k), \tag{2.41}$$

respectively, where r is fixed, say $r = 5$. Clearly, for $t \geq T$ and moderate k and s, we may expect that $B_t^+(k, r) \approx B_t^-(k, r)$, provided the variance of $\{\widehat{\gamma}_{2t}\}$ is bounded.

With this at hand, we define the stopping criterion as

$$T = \min \left\{ t \; : \; \frac{B_t^+(k, r) - B_t^-(k, r)}{B_t^-(k, r)} \leq \varepsilon \right\}, \tag{2.42}$$

where k and r are fixed and ε is a small number, say $\varepsilon \leq 0.01$.

Figure 2.7 displays the evolution of the the worst of the elite samples $(\widehat{\gamma}_t)$ for both the deterministic and noisy case, denoted by $\widehat{\gamma}_{1t}$ and $\widehat{\gamma}_{2t}$, respectively. We see in both cases a similar rapid drop in the level $\widehat{\gamma}_t$. It is important to note, however, that even though the algorithm in both the deterministic and noisy case converges to the optimal solution, the $\{\widehat{\gamma}_{2t}\}$ for the noisy case do not converge to γ^*, in contrast

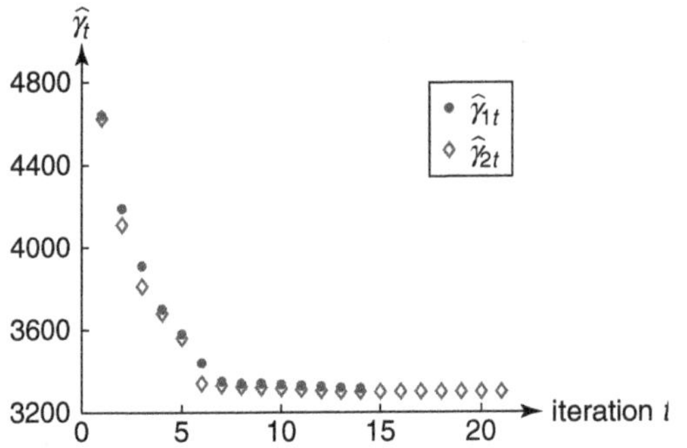

Figure 2.7 Evolution of the worst of the elite samples for a deterministic and noisy TSP.

to the $\{\widehat{\gamma}_{1t}\}$ for the deterministic case. This is because the latter eventually estimates the $(1-\rho)$-quantile of the deterministic $S(\boldsymbol{x}^*)$, whereas the former estimates the $(1-\rho)$-quantile of $\widehat{S}(\boldsymbol{x}^*)$, which is random. To estimate $S(\boldsymbol{x}^*)$ in the noisy case, one needs to take the sample average of $\widehat{S}(\boldsymbol{x}_T)$, where $\boldsymbol{x}_T$ is the solution found at the final iteration. □

Chapter 3

Minimum Cross-Entropy Method

This chapter deals with the *minimum cross-entropy* method, also known as the *MinxEnt* method for combinatorial optimization problems and rare-event probability estimation. The main idea of MinxEnt is to associate with each original optimization problem an auxiliary single-constrained convex optimization program in terms of probability density functions. The beauty is that this auxiliary program has a closed-form solution, which becomes the optimal zero variance solution, provided the "temperature" parameter is set to minus infinity. In addition, the associated pdf based on the product of marginals obtained from the joint optimal zero variance pdf coincide with the parametric pdf of the cross-entropy (CE) method. Thus, we obtain a strong connection between CE and MinxEnt, providing solid mathematical foundations.

3.1 INTRODUCTION

Let $H(\boldsymbol{x})$ be a continuous function defined on some closed bounded n-dimensional domain $\mathcal{X}$. Assume that $\boldsymbol{x}^*$ is a unique minimum point over $\mathcal{X}$. The following theorem is due to Pincus [94].

Theorem 3.1

Let $H(\boldsymbol{x})$ be a real-valued continuous function over closed bounded n-dimensional domain $\mathcal{X}$. Further assume that there is a unique minimum point $\boldsymbol{x}^*$ over $\mathcal{X}$ at which $\min_x H(\boldsymbol{x})$ *attains (there is no restriction on the number of local minima). Then the coordinates of x_k^*, $k = 1, \ldots, n$ of $\boldsymbol{x}^*$ are given by*

$$x_k^* = \lim_{\lambda \to \infty} \frac{\int_{\mathcal{X}} x_k \exp\left(-\lambda H(\boldsymbol{x})\right) d\boldsymbol{x}}{\int_{\mathcal{X}} \exp\left(-\lambda H(\boldsymbol{x})\right) d\boldsymbol{x}}, \quad k = 1, \ldots, n. \tag{3.1}$$

Fast Sequential Monte Carlo Methods for Counting and Optimization, First Edition.
Reuven Y. Rubinstein, Ad Ridder, and Radislav Vaisman.

The proof of the theorem is based on Laplace's formula, which for sufficiently large λ can be written as

$$\int_{\mathcal{X}} x_k \exp(-\lambda H(\boldsymbol{x}))\, d\boldsymbol{x} \approx x_k^* \exp(-\lambda H(\boldsymbol{x}^*)),$$

$$\int_{\mathcal{X}} \exp(-\lambda H(\boldsymbol{x}))\, d\boldsymbol{x} \approx \exp(-\lambda H(\boldsymbol{x}^*)).$$

This is because, for large λ, the major contribution to the integrals appearing in (3.1) comes from a small neighborhood of the minimizer $\boldsymbol{x}^*$.

Pincus' theorem holds for discrete optimization as well (assuming $|\mathcal{X}| < \infty$). In this case, the integrals should be replaced by the relevant sums.

There are many Monte Carlo methods for evaluating the coordinates of $\boldsymbol{x}^*$, that is, for approximating the ratio appearing in (3.1). Among them is the celebrated simulated annealing method, which is based on a Markov chain Monte Carlo technique, also called Metropolis' sampling procedure. The idea of the method is to sample from the Boltzmann density

$$g(\boldsymbol{x}) = \frac{\exp(-\lambda H(\boldsymbol{x}))}{\int_{\mathcal{X}} \exp(-\lambda H(\boldsymbol{x}))\, d\boldsymbol{x}} \tag{3.2}$$

without resorting to calculation of the integral (the denumerator). For details see [108].

It is important to note that, in general, sampling from the complex multidimensional pdf $g(\boldsymbol{x})$ is a formidable task. If, however, the function $H(\boldsymbol{x})$ is separable, that is, it can be represented by

$$H(\boldsymbol{x}) = \sum_{k=1}^{n} H_k(x_k),$$

then the pdf $g(\boldsymbol{x})$ in (3.2) decomposes as the product of its marginal pdfs, that is, it can be written as

$$g(\boldsymbol{x}) = \frac{\prod_{k=1}^{n} \exp(-\lambda H_k(x_k))}{\prod_{k=1}^{n} \int_{\mathcal{X}} \exp(-\lambda H_k(x_k))\, dx_k}. \tag{3.3}$$

Clearly, for a decomposable function $H(\boldsymbol{x})$ sampling from the one-dimensional marginal pdfs of $g(\boldsymbol{x})$ is fast.

Consider the application of the simulated annealing method to combinatorial optimization problems. As an example, consider the traveling salesman problem with n cities. In this case, [1] shows how simulated annealing runs a Markov chain $\boldsymbol{Y}$ with $(n-1)!$ states and $H(\boldsymbol{x})$ denoting the length of the tour. As $\lambda \to \infty$ the stationary distribution of $\boldsymbol{Y}$ will become a degenerated one, that is, it converges to the optimal solution $\boldsymbol{x}^*$ (shortest tour in the case of traveling salesman problem). It can be proved [1] that, in the case of multiple solutions, say R solutions, we have that as $\lambda \to \infty$ the stationary distribution of $\boldsymbol{Y}$ will be uniform on the set of the R optimal solutions.

The main drawback of simulated annealing is that it is slow and λ, called the annealing temperature, must be chosen heuristically.

We present a different Monte Carlo method, which we call MinxEnt. It is also associated with the Boltzmann distribution, however, it is obtained by solving a MinxEnt program of a special type and is suitable for rare-event probability estimation and approximation of the optimal solutions of a broad class of NP-hard linear integer and combinatorial optimization problems.

The main idea of the MinxEnt approach is to associate with each original optimization problem an auxiliary single-constrained convex MinxEnt program of a special type, which has a closed form solution.

The rest of this chapter is organized as follows. In Section 3.2, we present some background on the classic MinxEnt program. In Section 3.3, we establish connections between rare-event probability estimation and MinxEnt, and present what is called the basic MinxEnt (BME). Section 3.4 presents a new MinxEnt method, which involves indicator functions and is called the indicator MinxEnt (IME). We prove that the optimal pdf obtained from the solution of the IME program is a zero variance one, provided the temperature parameter is set to minus infinity. This is quite a remarkable result. In addition, we prove that the parametric pdf based on the product of marginals obtained from the optimal zero-variance pdf coincides with the parametric pdf of the standard CE method, which we covered in the previous chapter. A remarkable consequence is that we obtain a strong mathematical foundation for CE. In Section 3.5 we present the IME algorithm for optimization.

3.2 CLASSIC MinxEnt METHOD

The classic MinxEnt program reads as

$$\min_g \mathcal{D}(g,h) = \min_g \int \log \frac{g(\boldsymbol{x})}{h(\boldsymbol{x})}\, g(\boldsymbol{x})\, d\boldsymbol{x} = \min_g \mathbb{E}_g\left[\log \frac{g(\boldsymbol{X})}{h(\boldsymbol{X})}\right]$$

$$\text{s.t.} \int S_i(\boldsymbol{x}) g(\boldsymbol{x})\, d\boldsymbol{x} = \mathbb{E}_g[S_i(\boldsymbol{X})] = b_i, \quad \text{for all } i = 1,\ldots,m \tag{3.4}$$

$$\int g(\boldsymbol{x})\, d\boldsymbol{x} = 1.$$

Here, g and h are n-dimensional pdfs; $S_i(\boldsymbol{x}),\ i = 1,\ldots,m$ are given functions; and $\boldsymbol{x}$ is an n-dimensional vector. Here, h is assumed to be known and is called the prior pdf. The program (3.4) is called the classic minimum cross-entropy or simply the MinxEnt program. If the prior h is constant, then $\mathcal{D}(g,h) = \int g(\boldsymbol{x}) \log\, g(\boldsymbol{x})\, d\boldsymbol{x} +$ constant, so that the minimization of $\mathcal{D}(g,h)$ in (3.4) can be replaced with the maximization of

$$\mathcal{S}(g) = -\int g(\boldsymbol{x}) \log\, g(\boldsymbol{x})\, d\boldsymbol{x} = -\mathbb{E}_g[\log\, g(\boldsymbol{X})], \tag{3.5}$$

where $\mathcal{S}(g)$ is the Shannon entropy [63]. The corresponding program is called the Jaynes' MinxEnt program. Note that the former minimizes the Kullback-Leibler cross-entropy, while the latter maximizes the Shannon entropy [63]. For a good paper on the generalization of MinxEnt, see [16].

The MinxEnt program, which under mild conditions [8] presents a convex constrained functional optimization problem, can be solved via Lagrange multipliers. The solution is given by [8]

$$g(\boldsymbol{x}) = \frac{h(\boldsymbol{x}) \exp\left(-\sum_{i=1}^{m} \lambda_i S_i(\boldsymbol{x})\right)}{\mathbb{E}_h\left[\exp\left(-\sum_{i=1}^{m} \lambda_i S_i(\boldsymbol{X})\right)\right]}, \tag{3.6}$$

where $\lambda_i, i = 1, \ldots, m$ are obtained from the solution of the following system of equations:

$$\frac{\mathbb{E}_h\left[S_i(\boldsymbol{X}) \exp\left(-\sum_{j=1}^{m} \lambda_j S_j(\boldsymbol{X})\right)\right]}{\mathbb{E}_h\left[\exp\left(-\sum_{j=1}^{m_j} \lambda_j S_j(\boldsymbol{X})\right)\right]} = b_i. \tag{3.7}$$

The MinxEnt solution $g(\boldsymbol{x})$ can be written as

$$g(\boldsymbol{x}) = C(\lambda_1, \ldots, \lambda_m) h(\boldsymbol{x}) \exp\left(-\sum_{i=1}^{m} \lambda_i S_i(\boldsymbol{x})\right), \tag{3.8}$$

where

$$C^{-1}(\lambda_1, \ldots, \lambda_m) = \mathbb{E}_h\left[\exp\left(-\sum_{i=1}^{m} \lambda_i S_i(\boldsymbol{X})\right)\right] \tag{3.9}$$

is the normalization constant.

In the particular case where each $S_i(\boldsymbol{X}),\ \boldsymbol{X} = (X_1, \ldots, X_n)$ is coordinate-wise separable, that is,

$$S_i(\boldsymbol{X}) = \sum_{k=1}^{n} S_{ik}(X_k), \quad i = 1, \ldots, m \tag{3.10}$$

and the components X_k, $k = 1, \ldots, n$ of the random vector $\boldsymbol{X} = (X_1, \ldots, X_n)$ are independent. The joint pdf $g(\boldsymbol{x})$ in (3.6) reduces to the product of marginal pdfs. In such case, we say that $g(\boldsymbol{x})$ is decomposable.

In particular, the k-th component of $g(\boldsymbol{x})$ can be written as

$$g_k(x_k) = \frac{h_k(x) \exp\left(-\sum_{i=1}^{m} \lambda_i S_{ik}(x_k)\right)}{\mathbb{E}_{h_k}\left[\exp\left(-\sum_{i=1}^{m} \lambda_i S_{ik}(X_k)\right)\right]}, \quad k = 1, \ldots, n. \tag{3.11}$$

Remark 3.1

It is well known [34] that the optimal solution of the single-dimensional single-constrained MinxEnt program

$$
\begin{aligned}
&\min_g \mathcal{D}(g,h) = \min_g \mathbb{E}_g\left[\log \frac{g(\boldsymbol{X})}{h(\boldsymbol{X})}\right] \\
&\text{s.t. } \mathbb{E}_g[S(\boldsymbol{X})] = b, \\
&\int g(\boldsymbol{x})\, d\boldsymbol{x} = 1
\end{aligned}
\tag{3.12}
$$

coincides with the celebrated optimal exponential change of measure (ECM). Note that typically in a multidimensional ECM, one twists each component separately, using possibly different twisting parameters. In contrast, the optimal solution to the MinxEnt program is parameterized by a single-dimensional parameter λ, so for the multidimensional case ECM differs from MinxEnt.

EXAMPLE 3.1 ***Die Tossing***

To obtain better insight into the MinxEnt program, consider a particular case of (3.12) associated with die tossing. We assume $S(x) = x$ and $h(x) = f(x; \boldsymbol{u})$ is a given discrete distribution (the prior) over the six faces of the die, where $\boldsymbol{u} = (u_1, \ldots, u_6)$ denotes the nominal parameter vector. We restrict to densities $f(x; \boldsymbol{v})$ that are parameterized in this way; that is, $\boldsymbol{v} = (v_1, \ldots, v_6)$ with $v_k \geq 0$ for all k, such that $\sum_{k=1}^{6} v_k = 1$. Then we denote the Kullback-Leibler cross-entropy by $\mathcal{D}(\boldsymbol{v}, \boldsymbol{u})$, which stands for $\mathcal{D}(f(\cdot; \boldsymbol{v}), f(\cdot; \boldsymbol{v}))$. Similarly, the Shannon entropy $\mathcal{S}(f(\cdot; \boldsymbol{v}))$ is abbreviated to $\mathcal{S}(\boldsymbol{v})$.

In this die problem it is readily seen that the functional program (3.12) leads to the following parametric one:

$$
\begin{aligned}
&\min_{\boldsymbol{v}} \mathcal{D}(\boldsymbol{v}, \boldsymbol{u}) = \min_{\boldsymbol{v}} \sum_{k=1}^{6} v_k \log \frac{v_k}{u_k} \\
&\text{s.t. } \sum_{k=1}^{6} k v_k = b, \\
&\sum_{k=1}^{6} v_k = 1.
\end{aligned}
\tag{3.13}
$$

The optimal parameter vector $\boldsymbol{v}^* = (v_1, \ldots, v_6)$, derived from the solution of (3.13), can be written component-wise as

$$
v_k = \frac{u_k \exp(-k\lambda)}{\sum_{r=1}^{6} u_r \exp(-r\lambda)} = \frac{\mathbb{E}_{\boldsymbol{u}}[I_{\{X=k\}} \exp(-X\lambda)]}{\mathbb{E}_{\boldsymbol{u}}[\exp(-X\lambda)]}, \quad k = 1, \ldots, 6, \tag{3.14}
$$

Table 3.1 λ, $\boldsymbol{v}$, and $\mathcal{S}(\boldsymbol{v})$ as Function of b for a Fair Die

b	λ	v_1	v_2	v_3	v_4	v_5	v_6	$\mathcal{S}(\boldsymbol{v})$
1.0	∞	1.0000	0.0000	0.0000	0.0000	0.0000	0.0000	0.00000
1.5	1.0870	0.6637	0.2238	0.0755	0.0255	0.0086	0.0029	0.95356
2.0	0.6296	0.4781	0.2548	0.1357	0.0723	0.0385	0.0205	1.36724
2.5	0.3710	0.3475	0.2398	0.1654	0.1142	0.0788	0.0544	1.61373
3.0	0.1746	0.2468	0.2072	0.1740	0.1461	0.1227	0.1031	1.74843
3.5	0.0000	0.1666	0.1666	0.1666	0.1666	0.1666	0.1666	1.79176
4.0	−0.1746	0.1031	0.1227	0.1461	0.1740	0.2072	0.2468	1.74843
4.5	−0.3710	0.0544	0.0788	0.1142	0.1654	0.2398	0.3475	1.61373
5.0	−0.6296	0.0205	0.0385	0.0723	0.1357	0.2548	0.4781	1.36724
5.5	−1.0870	0.0029	0.0086	0.0255	0.0755	0.2238	0.6637	0.95356
6.0	$-\infty$	0.0000	0.0000	0.0000	0.0000	0.0000	1.0000	0.00000

where λ is derived from the numerical solution of

$$\frac{\sum_{k=1}^{6} k u_k \exp\,(-k\lambda)}{\sum_{k=1}^{6} u_k \exp\,(-k\lambda)} = b. \tag{3.15}$$

Table 3.1, which is an exact replica of Table 4.1 of [63], presents λ, $\boldsymbol{v}$, and the entropy $\mathcal{S}(\boldsymbol{v})$ as functions of b for a fair die, that is, with the prior ($u_1 = \frac{1}{6}, \ldots, u_6 = \frac{1}{6}$). The table is self-explanatory.

Note that

- If $b = 3.5$, then $\boldsymbol{v} = \boldsymbol{u} = (\frac{1}{6}, \ldots, \frac{1}{6})$ and, thus, $g = h$.
- $\mathcal{S}(\boldsymbol{v})$ is strictly concave in b and the maximal entropy $\max_b \mathcal{S}(\boldsymbol{v}) = \mathcal{S}(3.5) = 1.79176$.
- For the extreme values of b, that is, for $b = 6$ and $b = 1$, the corresponding optimal solutions are $\boldsymbol{v}^* = (0, 0, \ldots, 1)$ and $\boldsymbol{v}^* = (1, 0, \ldots, 0)$, respectively; that is, the pdf g becomes degenerated. For these cases:
  1. The entropy is $\mathcal{S}(\boldsymbol{v}) = 0$, and thus there is no uncertainty (for both degenerated vectors, $\boldsymbol{v} = (0, 0, \ldots, 1)$ and $\boldsymbol{v} = (1, 0, \ldots, 0)$).
  2. For $\boldsymbol{v} = (0, 0, \ldots, 1)$ and $\boldsymbol{v} = (1, 0, \ldots, 0)$ we have that $\lambda = -\infty$ and $\lambda = \infty$, respectively. This important observation is in the spirit of Pincus [94] Theorem 3.1 and will play an important role below.
  3. It can also be readily shown that $\boldsymbol{v}$ is degenerated regardless of the prior $\boldsymbol{u}$. □

The above observations for the die example can be readily extended to the case where instead of $S(x) = x$ one considers $S(x) = \sum_{k=1}^{r} a_k I_{\{x=k\}}$ with $r > 1$ and with arbitrary a_k's.

Remark 3.2

Taking into account that MaxEnt (with the objective function $\mathcal{S}(g)$, see (3.5)) can be viewed as a particular case of MinxEnt with constant prior $h(\boldsymbol{x})$, we can rewrite the basic MinxEnt formulas (3.6) and (3.7) as

$$g(\boldsymbol{x}) = \frac{\exp\left(-\sum_{i=1}^{m} \lambda_i S_i(\boldsymbol{x})\right)}{\int \exp\left(-\sum_{i=1}^{m} S_i(\boldsymbol{x})\lambda_i\right)\, d\boldsymbol{x}}, \tag{3.16}$$

where λ_i, $i = 1, \ldots, m$ are obtained from the solution of the following system of equations:

$$\frac{\int S_i(\boldsymbol{x}) \exp\left(-\sum_{j=1}^{m} \lambda_j S_j(\boldsymbol{x})\right)\, d\boldsymbol{x}}{\int \exp\left(-\lambda \sum_{j=1}^{m_j} S_j(\boldsymbol{x})\right)\, d\boldsymbol{x}} = b_i. \tag{3.17}$$

We extend next the MinxEnt program (3.4) to both equality and inequality constraints; that is, we consider the following general MinxEnt program:

$$\begin{aligned}
&\min_g \mathcal{D}(g,h) = \min_g \int \log \frac{g(\boldsymbol{x})}{h(\boldsymbol{x})}\, g(\boldsymbol{x})\, d\boldsymbol{x} = \min_g \mathbb{E}_g\left[\log \frac{g(\boldsymbol{X})}{h(\boldsymbol{X})}\right] \\
\text{s.t.}\ &\int S_i(\boldsymbol{x}) g(\boldsymbol{x})\, d\boldsymbol{x} = \mathbb{E}_g[S_i(\boldsymbol{X})] = b_i, \quad \text{for all } i = 1, \ldots, m_1 \\
&\int S_i(\boldsymbol{x}) g(\boldsymbol{x})\, d\boldsymbol{x} = \mathbb{E}_g[S_i(\boldsymbol{X})] \geq b_i, \quad \text{for all } i = m_1 + 1, \ldots, m_1 + m_2 \\
&\int g(\boldsymbol{x})\, d\boldsymbol{x} = 1.
\end{aligned} \tag{3.18}$$

In this case, applying the Kuhn-Tucker [75] conditions to the program (3.18) we readily obtain that $g(\boldsymbol{x})$ remains the same as in (3.6), while $\boldsymbol{\lambda} = (\lambda_0, \lambda_1, \ldots, \lambda_m)$, $m = m_1 + m_2$ are found from the solution of the following convex program:

$$\begin{aligned}
&\max_{\boldsymbol{\lambda}} \left(-\sum_{i=0}^{m} \lambda_i b_i - \mathbb{E}_h\left[\exp\left(-\sum_{i=0}^{m} \lambda_i S_i(\boldsymbol{X})\right)\right]\right) \\
&\text{s.t. } \lambda_i \geq 0, \quad \text{for all } i = m_1 + 1, \ldots, m.
\end{aligned} \tag{3.19}$$

3.3 RARE EVENTS AND MinxEnt

To establish the connection between MinxEnt and rare events, we consider the problem of computing a rare-event probability of the form

$$\ell = \mathbb{P}_h(S(\boldsymbol{X}) \geq b) = \mathbb{E}_h\left[I_{\{S(\boldsymbol{X}) \geq b\}}\right]. \tag{3.20}$$

Here, $\boldsymbol{X} \in \mathbb{R}^n$ is a random vector with pdf $h(\cdot)$, $S : \mathbb{R}^n \to \mathbb{R}$ is a performance function, and b is such that $\{S(\boldsymbol{X}) \geq b\}$ is a rare event.

Suppose that we estimate ℓ by the importance sampling estimator

$$\widehat{\ell} = \frac{1}{N}\sum_{k=1}^{N} I_{\{S(\boldsymbol{X}_k)\geq b\}}\; W(\boldsymbol{X}_k; h, g), \tag{3.21}$$

using the importance sampling pdf g (see Section 2.2). Then the MinxEnt program that determines g has the following single-constrained formulation:

$$\begin{aligned}
&\min_g \mathcal{D}(g,h) = \min_g \int \log \frac{g(\boldsymbol{x})}{h(\boldsymbol{x})}\, g(\boldsymbol{x})\, d\boldsymbol{x} = \min_g \mathbb{E}_g\left[\log \frac{g(\boldsymbol{X})}{h(\boldsymbol{X})}\right] \\
&\text{s.t. } \mathbb{E}_g[S(\boldsymbol{X})] = b, \\
&\int g(\boldsymbol{x})\, d\boldsymbol{x} = 1.
\end{aligned} \tag{3.22}$$

The solution is (cf. (3.6))

$$g(\boldsymbol{x}) = \frac{h(\boldsymbol{x}) \exp\left(-\lambda S(\boldsymbol{x})\right)}{\mathbb{E}_h[\exp\left(-\lambda S(\boldsymbol{X})\right)]}, \tag{3.23}$$

and λ solves

$$\frac{\mathbb{E}_h[S(\boldsymbol{X}) \exp\left(-\lambda S(\boldsymbol{X})\right)]}{\mathbb{E}_h[\exp\left(-\lambda S(\boldsymbol{X})\right)]} = b. \tag{3.24}$$

We call (3.22) and (3.23), (3.24) the basic MinxEnt (BME) program and solution, respectively. This is the nonparametric setting in which generally the solution $g(\boldsymbol{x})$ is a difficult pdf to generate samples from. As a typical example, suppose that the prior $h(\boldsymbol{x})$ is the uniform pdf on a bounded state space $\mathcal{X} \subset \mathbb{R}^n$, that is $h(\boldsymbol{x}) = 1/|\mathcal{X}| I_{\{\boldsymbol{x}\in\mathcal{X}\}}$, where $|\mathcal{X}|$ denotes the total area of the set $\mathcal{X}$. The rare event is $\mathcal{X}^* = \{S(\boldsymbol{x}) \geq b\} \subset \mathcal{X}$. The importance sampling estimator (3.21) using the MinxEnt solution (3.23) becomes

$$\widehat{\ell} = \frac{1}{N}\sum_{k=1}^{N} I_{\{S(\boldsymbol{X}_k)\geq b\}} \frac{h(\boldsymbol{X}_k)}{g(\boldsymbol{X}_k)} = \frac{1}{N}\sum_{k=1}^{N} I_{\{S(\boldsymbol{X}_k)\geq b\}} \frac{1}{g(\boldsymbol{X}_k)}\frac{1}{|\mathcal{X}|},$$

where $\boldsymbol{X}_1, \ldots, \boldsymbol{X}_N$ is a random sample from g.

We give two simple examples; the first one leads to a simple simulation procedure from g, the second one does not.

EXAMPLE 3.2

Suppose that $\mathcal{X} = [0, 1]^2$, $S(x_1, x_2) = x_1 + x_2$, and $b = 2 - \sqrt{2\varepsilon}$. In this case $h(x_1, x_2) = 1$ on $\mathcal{X}$, and

$$\ell = \mathbb{P}_h(X_1 + X_2 > b) = \varepsilon,$$

where X_1 and X_2 are independent uniform $U(0, 1)$ random variables. Thus,

$$\begin{aligned}\mathbb{E}_h[\exp(-\lambda S(\boldsymbol{X}))] &= \mathbb{E}_h[\exp(-\lambda(X_1 + X_2))] \\ &= \mathbb{E}_h[\exp(-\lambda X_1)]\,\mathbb{E}_h[\exp(-\lambda X_2)] \\ &= \left(\int_0^1 e^{-\lambda x}\,dx\right)^2 = \left(\frac{1 - e^{-\lambda}}{\lambda}\right)^2.\end{aligned}$$

The parameter λ solves

$$\frac{d}{d\lambda}\log \mathbb{E}_h[\exp(-\lambda S(\boldsymbol{X}))] = b \quad \Leftrightarrow \quad \frac{e^{-\lambda}}{1 - e^{-\lambda}} - \frac{1}{\lambda} = 1 - \frac{1}{2}\sqrt{2\varepsilon}.$$

This is easily solved numerically for any given ε. The importance sampling pdf becomes

$$g(x_1, x_2) = \frac{h(x_1, x_2)e^{-\lambda(x_1+x_2)}}{\mathbb{E}_h[e^{-\lambda(X_1+X_2)}]} = \frac{\lambda e^{-\lambda x_1}}{1 - e^{-\lambda}} \cdot \frac{\lambda e^{-\lambda x_2}}{1 - e^{-\lambda}}$$

on $\mathcal{X}$, otherwise $g(x_1, x_2) = 0$. Hence, under the importance sampling pdf g, X_1 and X_2 are again independent but have a conditional exponential distribution with parameter λ, conditioned on $x \in [0, 1]$. Generating samples from g is straightforward. □

EXAMPLE 3.3

Suppose that $\mathcal{X} = \{\boldsymbol{x} = (x_1, x_2) : x_1^2 + x_2^2 \leq 2\}$, $S(x_1, x_2) = x_1 + x_2$, and $b = 2 - \varepsilon$. Thus, $h(x_1, x_2) = 1/(2\pi)$ on $\mathcal{X}$, and

$$\ell = \mathbb{P}_h(X_1 + X_2 > b) = \frac{B(\varepsilon)}{2\pi},$$

where $B(\varepsilon)$ is the circle segment above the chord $x_1 + x - 2 = b$. Clearly, $\ell \to 0$ when $\varepsilon \to 0$. Note that X_1 and X_2 are dependent. Generating samples from h is easy and can be done, for example, by acceptance-rejection from uniform samples in the square $[-\sqrt{2}, \sqrt{2}] \times [-\sqrt{2}, \sqrt{2}]$. In this example we get explicit expressions neither for the normalizing constant $\mathbb{E}_h[\exp(-\lambda(X_1 + X_2))]$ nor for the importance sampling pdf g. □

As we already mentioned in Chapter 2, the cross-entropy method assumes that the prior pdf $h(\cdot) = f(\cdot; \boldsymbol{u})$ belongs to a parameterized family $\mathscr{F}$ of pdfs (2.6), where $\boldsymbol{u}$ is called the nominal parameter vector. The solution $g(\cdot)$ to the MinxEnt program as given in (3.23) is generally not a pdf in that parameterized family. In some cases it is possible to find an approximate pdf $f(\cdot; \boldsymbol{v})$ as in the following setting.

Definition 3.1 ***Marginal Expectation Parameterization***

We say that the parameterized family $\mathscr{F}$ of pdfs satisfies marginal expectation parameterization when

$$v_j = \mathbb{E}_{\boldsymbol{v}}[X_j] \text{ for all } j = 1, \ldots, n.$$

Let $h(\boldsymbol{x}) = f(\boldsymbol{x}; \boldsymbol{u}) \in \mathscr{F}$ be the prior pdf. The MinxEnt solution for $\mathbb{E}_g[X_j]$ is

$$\mathbb{E}_g[X_j] = \frac{\mathbb{E}_{\boldsymbol{u}}[X_j \exp(-\lambda S(\boldsymbol{X}))]}{\mathbb{E}_{\boldsymbol{u}}[\exp(-\lambda S(\boldsymbol{X}))]}, \quad j = 1, \ldots, n. \tag{3.25}$$

Suppose that the family $\mathscr{F}$ satisfies the marginal expectation parameterization, and let $\boldsymbol{v} = (v_1, \ldots, v_n)$ be any parameter vector for the family $\mathscr{F}$. Then the above analysis suggests carrying out the importance sampling using v_j equal to $\mathbb{E}_g[X_j]$ given in (3.25). Hence, we approximate the MinxEnt solution $g(\cdot)$ by $f(\cdot; \boldsymbol{v})$ with

$$v_j = \frac{\mathbb{E}_{\boldsymbol{u}}[X_j \exp(-\lambda S(\boldsymbol{X}))]}{\mathbb{E}_{\boldsymbol{u}}[\exp(-\lambda S(\boldsymbol{X}))]}, \quad j = 1, \ldots, n. \tag{3.26}$$

Note that formula (3.26) is similar to the corresponding cross-entropy solution [107]

$$v_j = \frac{\mathbb{E}_{\boldsymbol{u}}\left[X_j \, I_{\{S(\boldsymbol{X}) \geq b\}}\right]}{\mathbb{E}_{\boldsymbol{u}}\left[I_{\{S(\boldsymbol{X}) \geq b\}}\right]}, \tag{3.27}$$

with one main difference: the indicator function $I_{\{S(\boldsymbol{X}) \geq b\}}$ in the CE formula is replaced by $\exp(-\lambda \, S(\boldsymbol{X}))$.

EXAMPLE 3.4

Let $n = 6$, and suppose that $\boldsymbol{X} = (X_1, \ldots, X_6) \in \{0, 1\}^6$ is a binary random vector with independent components and probabilities

$$\boldsymbol{u} = (u_1, \ldots, u_6) = (0.003, 0.005, 0.002, 0.001, 0.005, 0.003).$$

In other words,

$$\boldsymbol{X} \sim f(\boldsymbol{x}; \boldsymbol{u}) = \prod_{j=1}^{6} u_j^{x_j} (1 - u_j)^{1-x_j}, \quad \boldsymbol{x} \in \{0, 1\}^6.$$

Assume further that the performance function is $S(\boldsymbol{x}) = \sum_{j=1}^{6} x_j$, and we want to find the optimal parameter $\boldsymbol{v}$ to calculate the probability of the event $\{S(\boldsymbol{x}) \geq 4\}$.

Because $S(\boldsymbol{x})$ is decomposable, the MinxEnt program results in a product form

$$f(\boldsymbol{x}; \boldsymbol{v}) = \prod_{j=1}^{6} f_j(x_j; v_j) = \prod_{j=1}^{6} v_j^{x_j} (1 - v_j)^{1-x_j},$$

with

$$\begin{aligned}
f(\boldsymbol{x};\boldsymbol{v}) &= \frac{f(\boldsymbol{x};\boldsymbol{u})e^{-\lambda S(\boldsymbol{x})}}{\mathbb{E}_{\boldsymbol{u}}[e^{-\lambda S(\boldsymbol{X})}]} \\
\Leftrightarrow \quad \prod_{j=1}^{6} f_j(x_j;v_j) &= \frac{\prod_{j=1}^{6} f_j(x_j;u_j)e^{-\lambda x_j}}{\prod_{j=1}^{6}\mathbb{E}_{u_j}[e^{-\lambda X_j}]} \\
\Leftrightarrow \quad v_j &= \frac{u_j e^{-\lambda}}{1-u_j+u_j e^{-\lambda}}, \qquad j=1,\ldots,6.
\end{aligned}$$

The parameter λ is obtained by solving

$$\frac{d}{d\lambda}\log\ \mathbb{E}_{\boldsymbol{u}}[\exp\,(-\lambda S(\boldsymbol{X}))] = b = 4 \quad \Leftrightarrow \quad \sum_{j=1}^{6}\frac{u_j e^{-\lambda}}{1-u_j+u_j e^{-\lambda}} = b.$$

This equation is solved numerically, resulting in $\lambda = -6.623455$ and

$$\boldsymbol{v} = (v_1,\ldots,v_6) = (0.694, 0.791, 0.601, 0.430, 0.791, 0.694).$$

The CE solution is given in (3.27), which is estimated by the adaptive CE algorithm. Note that in this simple example the MinxEnt program does not need to learn the optimal parameters, but it requires a numerical procedure to find them. We executed 100 simulation experiments of the CE and BME methods with different seeds and measured their average statistical and computing performance. The CE algorithm used a sample size $N = 5000$ and rarity parameter $\rho = 0.1$ at each iteration; the Lagrange parameter λ was found by bisection with an error of 10^{-10}. In the final importance sampling estimator, we used 2,000,000 samples. We found that under these assumptions the computing times were about the same, but the BME variance is smaller than the CE one: the reduction is 72%, where

$$\text{reduction factor} = 1 - \frac{\mathbb{V}\text{ar[BME]}}{\mathbb{V}\text{ar[CE]}}.$$

The gain is

$$\text{gain} = \frac{\mathbb{V}\text{ar[CE]} \times \text{CE computing time}}{\mathbb{V}\text{ar[BME]} \times \text{BME computing time}} \approx 1.25.$$

□

3.4 INDICATOR MinxEnt METHOD

Consider the set

$$\mathcal{X}^* = \{\boldsymbol{x} \in \mathbb{R}^n : S_i(\boldsymbol{x}) \geq b_i, \quad i = 1,\ldots,m\}, \tag{3.28}$$

where $S_i(\boldsymbol{x}),\quad i = 1,\ldots,m$ are arbitrary functions. As before, we assume that the random variable $\boldsymbol{X} \in \mathbb{R}^n$ is distributed according to a prior pdf $h(\cdot)$.

Hence, we associate with (3.28) the following multiple-event probability:

$$\ell = \mathbb{P}_h(\boldsymbol{X} \in \mathcal{X}^*) = \mathbb{P}_h\left(\bigcap_{i=1}^{m} \{S_i(\boldsymbol{X}) \geq b_i\}\right) = \mathbb{E}_h\left[\prod_{i=1}^{m} I_{\{S_i(\boldsymbol{X}) \geq b_i\}}\right]. \tag{3.29}$$

Note that (3.29) extends (3.20) in the sense that it involves simultaneously an intersection of m events $\{S_i(\boldsymbol{X}) \geq b_i\}$, that is, multiple events rather than a single one $\{S(\boldsymbol{X}) \geq \gamma\}$. Note also that some of the constraints may be equality ones, that is, $\{S_i(\boldsymbol{X}) = b_i\}$. Note finally that (3.29) has some interesting applications in rare-event simulation. For example, in a queueing model one might be interested in estimating the probability of the simultaneous occurrence of two events, $\{S_1(\boldsymbol{X}) \geq b_1\}$ and $\{S_2(\boldsymbol{X}) \geq b_2\}$, where the first is associated with buffer overflow (the number of customers S_1 is at least b_1), and the second is associated with the sojourn time (the waiting time of the customers S_2 in the queueing system is at least b_2).

We assume that each individual event $\{S_i(\boldsymbol{X}) \geq b_i\}$, $i = 1, \ldots, m$, is not rare, that is each probability $\mathbb{P}_h\{S_i(\boldsymbol{X}) \geq b_i\}$ is not a rare-event probability, say $\mathbb{P}_h\{S_i(\boldsymbol{X}) \geq b_i\} \geq 10^{-4}$, but their intersection forms a rare-event probability ℓ. Similar to the single-event case in (3.20) we are interested in efficient estimation of ℓ defined in (3.29).

The main idea of the indicator MinxEnt approach is to design an importance sampling pdf $g(\boldsymbol{x})$ such that under $g(\boldsymbol{x})$ all constraints $\{S_i(\boldsymbol{X}) \geq b_i,\ i = 1, \ldots, m\}$ are fulfilled with certainty. For that purpose, we define

$$C_i(\boldsymbol{X}) = I_{\{S_i(\boldsymbol{X}) \geq b_i\}} \quad \text{and} \quad \mathcal{C}(\boldsymbol{X}) = \sum_{i=1}^{m} C_i(\boldsymbol{X}). \tag{3.30}$$

Then the indicator MinxEnt program is defined by the following single-constrained MinxEnt program

$$\begin{aligned} &\min_g \mathcal{D}(g, h) = \min_g \mathbb{E}_g\left[\log \frac{g(\boldsymbol{X})}{h(\boldsymbol{X})}\right] \\ &\text{s.t. } \mathbb{E}_g[\mathcal{C}(\boldsymbol{X})] = m \\ &\int g(\boldsymbol{x})\, d\boldsymbol{x} = 1. \end{aligned} \tag{3.31}$$

Its solution $g(\boldsymbol{x})$ (see Section 3.2) is

$$g(\boldsymbol{x}) = \frac{h(\boldsymbol{x}) \exp\,(-\lambda \mathcal{C}(\boldsymbol{x}))}{\mathbb{E}_h[\exp\,(-\lambda \mathcal{C}(\boldsymbol{X}))]}, \tag{3.32}$$

where λ is obtained from the solution of the following equation:

$$\frac{\mathbb{E}_h[\mathcal{C}(\boldsymbol{X}) \exp\,(-\lambda \mathcal{C}(\boldsymbol{X}))]}{\mathbb{E}_h[\exp\,(-\lambda \mathcal{C}(X))} = m. \tag{3.33}$$

We call (3.31) and (3.32), (3.33) the indicator MinxEnt (IME) program and solution, respectively.

When (3.33) has no solution, then, in fact, the set $\{S_i(\boldsymbol{x}) \geq b_i, \quad i = 1, \dots, m\}$ is empty.

We will use $g(\boldsymbol{x})$ to estimate the rare-event probability ℓ given in (3.29). In forthcoming Lemmas 3.1 and 3.2 we will show that $g(\boldsymbol{x})$ coincides with the zero-variance optimal importance sampling pdf, that is, with

$$g(\boldsymbol{x}) = \frac{h(\boldsymbol{x}) I_{\{\boldsymbol{x} \in \mathcal{X}^*\}}}{\ell} = \frac{h(\boldsymbol{x}) I_{\{\mathcal{C}(\boldsymbol{x})=m\}}}{\mathbb{P}_h(\mathcal{C}(\boldsymbol{X}) = m)}. \tag{3.34}$$

Specifically, this gives

$$\mathbb{P}_g(\mathcal{C}(\boldsymbol{X}) = m) = \mathbb{E}_g\left[I_{\{\mathcal{C}(\boldsymbol{X})=m\}}\right] = 1. \tag{3.35}$$

Note that the classic multiconstrained MinxEnt program (3.4) involves expectations of $S_i(\boldsymbol{X})$, while the proposed single-constrained one (3.31) is based on the expectations of the indicators of $S_i(\boldsymbol{X})$, so the name indicator MinxEnt program or simply IME program.

For $m = 1$ the IME program (3.31) reduces to

$$\begin{aligned} &\min_g \ \mathcal{D}(g, h) = \min_g \mathbb{E}_g\left[\log \frac{g(\boldsymbol{X})}{h(\boldsymbol{X})}\right] \\ &\text{s.t. } \mathbb{E}_g[\mathcal{C}(\boldsymbol{X})] = 1 \\ &\int g(\boldsymbol{x}) \, d\boldsymbol{x} = 1, \end{aligned} \tag{3.36}$$

where $\mathcal{C}(\boldsymbol{X}) = I_{\{S(\boldsymbol{X}) \geq b\}}$.

Observe also that, in this case, the single-constrained programs (3.36) and (3.22) do not coincide: in the former case we use an expectation of the indicator of $S(\boldsymbol{X})$, that is, $\mathbb{E}[I_{\{S(\boldsymbol{X}) \geq b\}}]$, while in the latter case we use an expectation of $S(\boldsymbol{X})$, that is, $\mathbb{E}[S(\boldsymbol{X})]$. We shall treat the program (3.36) in more details in Section 3.5.

Lemma 3.1

The optimal λ *of the IME program* (3.31) *satisfying* (3.33) *is* $\lambda = -\infty$.

Proof

The proof is given for a discrete domain $\mathcal{X}$. For a continuous domain we replace the summations by integrations.

To prove that the optimal λ of the IME program (3.31) is $\lambda = -\infty$, we proceed as follows. Denoting, as before, $\mathcal{C}(\boldsymbol{x}) = \sum_{i=1}^{m} C_i(\boldsymbol{x}) \in \{0, 1, \ldots, m\}$ we can write (3.33) as

$$\begin{aligned}
&\lim_{\lambda\to-\infty} \frac{\mathbb{E}_h[\mathcal{C}(\boldsymbol{X})\exp\{-\lambda\mathcal{C}(\boldsymbol{X})\}]}{\mathbb{E}_h[\exp\{-\lambda\mathcal{C}(\boldsymbol{X})\}]} \\
&= \lim_{\lambda\to-\infty} \frac{\sum_{\boldsymbol{x}\in\mathcal{X}} h(\boldsymbol{x})\mathcal{C}(\boldsymbol{x})e^{-\lambda\mathcal{C}(\boldsymbol{x})}}{\sum_{\boldsymbol{x}\in\mathcal{X}} h(\boldsymbol{x})e^{-\lambda\mathcal{C}(\boldsymbol{x})}} \\
&= \lim_{\lambda\to-\infty} \frac{\sum_{\boldsymbol{x}\ \text{s.t.}\ \mathcal{C}(\boldsymbol{x})=m} h(\boldsymbol{x})\, m\, e^{-\lambda m} + \sum_{\boldsymbol{x}\ \text{s.t.}\ \mathcal{C}(\boldsymbol{x})<m} h(\boldsymbol{x})\mathcal{C}(\boldsymbol{x})e^{-\lambda\mathcal{C}(\boldsymbol{x})}}{\sum_{\boldsymbol{x}\ \text{s.t.}\ \mathcal{C}(\boldsymbol{x})=m} h(\boldsymbol{x})e^{-\lambda m} + \sum_{\boldsymbol{x}\ \text{s.t.}\ \mathcal{C}(\boldsymbol{x})<m} h(\boldsymbol{x})e^{-\lambda\mathcal{C}(\boldsymbol{x})}} \\
&= \lim_{\lambda\to-\infty} \frac{\sum_{\boldsymbol{x}\ \text{s.t.}\ \mathcal{C}(\boldsymbol{x})=m} h(\boldsymbol{x})\, m + \sum_{\boldsymbol{x}\ \text{s.t.}\ \mathcal{C}(\boldsymbol{x})<m} h(\boldsymbol{x})\mathcal{C}(\boldsymbol{x})e^{\lambda(m-\mathcal{C}(\boldsymbol{x}))}}{\sum_{\boldsymbol{x}\ \text{s.t.}\ \mathcal{C}(\boldsymbol{x})=m} h(\boldsymbol{x}) + \sum_{\boldsymbol{x}\ \text{s.t.}\ \mathcal{C}(\boldsymbol{x})<m} h(\boldsymbol{x})e^{\lambda(m-\mathcal{C}(\boldsymbol{x}))}} \\
&\overset{(i)}{=} \frac{\sum_{\boldsymbol{x}\ \text{s.t.}\ \mathcal{C}(\boldsymbol{x})=m} h(\boldsymbol{x})\, m}{\sum_{\boldsymbol{x}\ \text{s.t.}\ \mathcal{C}(\boldsymbol{x})=m} h(\boldsymbol{x})} = m,
\end{aligned}$$

where we used in equality (i) that $\lambda(m - \mathcal{C}(\boldsymbol{x})) \to -\infty$ when $\lambda \to -\infty$ because $m - \mathcal{C}(\boldsymbol{x}) > 0$. □

Lemma 3.2

For $\lambda = -\infty$ the optimal IME pdf $g(\boldsymbol{x})$ in (3.32) *corresponds to the zero-variance importance sampling pdf* (3.34).

Proof

The proof is given for a discrete domain $\mathcal{X}$. For a continuous domain we replace the summations by integrations. By Lemma 3.1, $\lambda \to -\infty$ and, denoting as before $\mathcal{C}(\boldsymbol{x}) = \sum_{i=1}^{m} C_i(\boldsymbol{x}) \in \{0, 1, \ldots, m\}$, we can write (3.32) as

$$\begin{aligned}
g(\boldsymbol{x}) &= \lim_{\lambda\to-\infty} \frac{h(\boldsymbol{x})e^{-\lambda\mathcal{C}(\boldsymbol{x})}}{\sum_{\boldsymbol{x}'\in\mathcal{X}} h(\boldsymbol{x}')e^{-\lambda\mathcal{C}(\boldsymbol{x}')}} \\
&= \lim_{\lambda\to-\infty} \frac{h(\boldsymbol{x})e^{-\lambda\mathcal{C}(\boldsymbol{x})}}{\sum_{\boldsymbol{x}'\ \text{s.t.}\ \mathcal{C}(\boldsymbol{x}')=m} h(\boldsymbol{x}')e^{-\lambda m} + \sum_{\boldsymbol{x}'\text{s.t.}\mathcal{C}(\boldsymbol{x}')<m} h(\boldsymbol{x}')e^{-\lambda\mathcal{C}(\boldsymbol{x}')}} \\
&= \lim_{\lambda\to-\infty} \frac{h(\boldsymbol{x})e^{\lambda(m-\mathcal{C}(\boldsymbol{x}))}}{\sum_{\boldsymbol{x}'\ \text{s.t.}\ \mathcal{C}(\boldsymbol{x}')=m} h(\boldsymbol{x}') + o(1)} \\
&= \lim_{\lambda\to-\infty} \frac{h(\boldsymbol{x})e^{\lambda(m-\mathcal{C}(\boldsymbol{x}))}}{\mathbb{P}_h(\mathcal{C}(\boldsymbol{X}) = m) + o(1)} \\
&= \begin{cases} 0, & \mathcal{C}(\boldsymbol{x}) \in \{0, 1, \ldots, m-1\}; \\ h(\boldsymbol{x})/\mathbb{P}_h(\mathcal{C}(\boldsymbol{X}) = m), & \mathcal{C}(\boldsymbol{x}) = m. \end{cases}
\end{aligned}$$

□

Observe again that generating samples from a multidimensional Boltzmann pdf, like $g(\boldsymbol{x})$ in (3.32), is a difficult task. One might apply Markov chain Monte Carlo algorithms [108], but these are time consuming.

Assume further that the prior $h(\boldsymbol{x})$ is given as a parameterized pdf $f(\boldsymbol{x};\boldsymbol{u})$. Then, similar to [103], we shall approximate $g(\boldsymbol{x})$ in (3.32) by a product of the marginal pdfs of a parameterized pdf $f(\boldsymbol{x};\boldsymbol{v})$. That is, $g_i(x_i) = f_i(x_i, v_i)$, $i = 1,\ldots,n$, with

$$v_i = \frac{\mathbb{E}_{\boldsymbol{u}}[X_i \exp(-\lambda \mathcal{C}(\boldsymbol{X}))]}{\mathbb{E}_{\boldsymbol{u}}[\exp(-\lambda \mathcal{C}(\boldsymbol{X}))]}, \quad i = 1,\ldots,n, \tag{3.37}$$

which coincides with (3.26) up to the notations. A further restriction assumes that each component of $\boldsymbol{X}$ is discrete random variable on $\{1,2,\ldots,r\}$. Then (3.37) extends to

$$\mathbb{P}_g(X_i = j) = v_{ij} = \frac{\mathbb{E}_{\boldsymbol{u}}[I_{\{X_i=j\}} \exp(-\lambda \mathcal{C}(\boldsymbol{X}))]}{\mathbb{E}_{\boldsymbol{u}}[\exp(-\lambda \mathcal{C}(\boldsymbol{X}))]}, \quad i = 1,\ldots,n; \quad j = 1,\ldots,r. \tag{3.38}$$

Remark 3.3 ***The Standard Cross-Entropy Method***

Similar to (3.37) (see also (3.27)) we can define the following CE updating formula:

$$v_j = \frac{\mathbb{E}_{\boldsymbol{u}}[X_j I_{\{C(\boldsymbol{X})=m\}}]}{\mathbb{E}_{\boldsymbol{u}}[I_{\{C(\boldsymbol{X})=m\}}]}. \tag{3.39}$$

3.4.1 Connection between CE and IME

To establish the connection between CE and IME we need the following result:

Theorem 3.2

For $\lambda = -\infty$ the optimal parameter vector $\boldsymbol{v}$ in (3.37) of the marginal pdfs of the optimal $g(\boldsymbol{x})$ in (3.32) coincides with the $\boldsymbol{v}$ in (3.39) for the CE method.

Proof

The proof is very similar to Lemma 3.2 and is omitted. □

Theorem 3.2 is crucial for the foundations of the CE method. Indeed, designed originally in [101] as a heuristics for rare-event estimation and combinatorial optimization problems, Theorem 3.2 states that CE has strong connections with the IME program (3.31). The main reason is that the optimal parametric pdf $f(\boldsymbol{x},\boldsymbol{v})$ (with $\boldsymbol{v}$ in (3.37) and $\lambda = -\infty$) and the CE pdf (with $\boldsymbol{v}$ as in (3.39)) obtained heuristically from the solution of the following cross-entropy program:

$$\min_{\boldsymbol{v}} \mathbb{E}_g\left[\log \frac{g(\boldsymbol{X})}{f(\boldsymbol{X};\boldsymbol{v})}\right]$$

are the same, provided $g(\boldsymbol{x})$ is the zero variance IS pdf.

The crucial difference between the proposed IME method and its CE counterparts lies in their simulation-based versions: in the latter we always require to generate a sequence of tuples $\{\boldsymbol{v}_t, m_t\}$, while in the former we can fix in advance the temperature parameter λ (to be set a large negative number) and then generate a sequence of parameter vectors $\{\boldsymbol{v}_t\}$ based on (3.37) alone. In addition, in contrast to CE, neither the elite sample nor the rarity parameter are involved in IME. As result, the proposed IME Algorithm becomes typically simpler, faster and at least as accurate as the standard CE based on (3.39).

3.5 IME METHOD FOR COMBINATORIAL OPTIMIZATION

We consider here unconstrained and constrained optimization, respectively.

3.5.1 Unconstrained Combinatorial Optimization

Consider the following non-smooth (continuous or discrete) unconstrained optimization program:

$$\max_{\boldsymbol{x} \in \mathbb{R}^n} S(\boldsymbol{x}).$$

Before proceeding with optimization recall that

1. The basic MinxEnt method of Section 3.3 is based on program (3.22), while the indicator MinxEnt method of Section 3.4 is based on program (3.36).
2. The programs (3.22) and (3.36) are different in the sense that, in the former, we require the condition $\mathbb{E}[S(\boldsymbol{X})] \geq b$, while for the latter we require the condition $\mathbb{E}[I_{\{S(\boldsymbol{X}) \geq b\}}] = 1$.
3. The parameter vector $\boldsymbol{v}$ in the basic MinxEnt is updated according to (3.26), where λ is obtained from (3.7), while in the indicator MinxEnt according to (3.37) (where $m = 1$).
4. The crucial difference between the two methods is that, in the basic MinxEnt, a sequence of the triplets $\{(\boldsymbol{v}_t, b_t, \lambda_t)\}$ is generated [103], while in the indicator MinxEnt only a sequence of tuples $\{(\boldsymbol{v}_t, b_t)\}$ is generated with λ being fixed and equal to a large negative number. The levels b_t are chosen again by considering the $1 - \rho$ quantile of the samples performance values, similar to the cross-entropy method.

If not stated otherwise, we shall use the indicator MinxEnt.

To proceed with $\max_{\boldsymbol{x} \in \mathbb{R}^n} S(\boldsymbol{x})$, denote by b^* the optimal function value. In this case, the indicator MinxEnt program becomes

$$\begin{aligned} &\min_g \mathcal{D}(g, h) = \min_g \mathbb{E}_g \left[\log \frac{g(\boldsymbol{X})}{f(\boldsymbol{x})} \right] \\ &\text{s.t. } \mathbb{E}_g[I_{\{S(\boldsymbol{X}) \geq b\}}] = 1 \end{aligned} \tag{3.40}$$

$$\int g(\boldsymbol{x})\, d\boldsymbol{x} = 1.$$

The corresponding updating of the component of the vector $\widehat{\boldsymbol{v}}_t$ can be written as

$$\widehat{v}_{t,j} = \frac{\sum_{k=1}^{N} X_{kj} \exp\left(-\lambda I_{\{S(\boldsymbol{X}_k) \geq \widehat{b}_t\}}\right)}{\sum_{k=1}^{N} \exp\left(-\lambda I_{\{S(\boldsymbol{X}_k) \geq \widehat{b}_t\}}\right)}, \tag{3.41}$$

where λ is a big negative number.

To motivate the program (3.40), consider again the die rolling example.

EXAMPLE 3.5 ***The Die Rolling Example Using Program* (3.40)**

Table 3.2 presents data similar to Table 3.1 for the die rolling example using the MinxEnt program (3.40) with $S(X) = X$. In particular, it presents $\boldsymbol{v}$ (updated according to (3.37)) and the entropy $\mathcal{S}(\boldsymbol{v})$ as functions of b for a fair die with the indicator of X, while calculating $\ell = \mathbb{P}(X \geq b)$ using (3.40) with $\lambda = -20$. One can see that, from the comparison of Table 3.2 and Table 3.1, that the entropy $\mathcal{S}(\boldsymbol{v})$ in the latter is smaller than in the former, which is based on the MinxEnt program (3.22). □

Algorithm 3.1 ***IME Algorithm for Unconstrained Optimization***

1. ***Initialize:*** *Define* $\widehat{\boldsymbol{v}}_0 = \boldsymbol{u}$, *choose* $f(\boldsymbol{x}; \boldsymbol{u})$ *uniformly distributed over* $\mathcal{X}$. *Set* λ *to a large negative number, say* $\lambda = -100$. *Set* $t = 1$ *(iteration = level counter).*
2. ***Draw:*** *Generate a sample* $\boldsymbol{X}_1, \ldots, \boldsymbol{X}_N$ *from the density* $f(\boldsymbol{x}; \widehat{\boldsymbol{v}}_{t-1})$.
3. ***Select:*** *Compute the elite sampling value* $\widehat{b}_t$ *of* $S(\boldsymbol{X}_1), \ldots, S(\boldsymbol{X}_N)$ *(similar to in CE).*
4. ***Update:*** *Use the same sample* $\boldsymbol{X}_1, \ldots, \boldsymbol{X}_N$ *and compute* $\widehat{\boldsymbol{v}}_t$, *according to* (3.41).

Table 3.2 $\boldsymbol{v}$ and $\mathcal{X}(\boldsymbol{v})$ as function of b for a Fair Die while Calculating $\ell = \mathbb{P}(X \geq b)$

b	v_1	v_2	v_3	v_4	v_5	v_6	$\mathcal{S}(\boldsymbol{v})$
1.0	0.17	0.17	0.17	0.17	0.17	0.17	1.7917
2.0	0.0	0.2	0.2	0.2	0.2	0.2	1.6094
3.0	0.0	0.0	0.25	0.25	0.25	0.25	1.3863
3.5	0.0	0.0	0.0	0.33	0.33	0.33	1.0485
4.0	0.0	0.0	0.0	0.33	0.33	0.33	1.0485
5.0	0.0	0.0	0.0	0.0	0.5	0.5	0.6931
6.0	0.0	0.0	0.0	0.0	0.0	1.0	0.0

5. ***Smooth:*** *Smooth out the vector* $\widehat{\boldsymbol{v}}_t$ *according to*

$$\widehat{\boldsymbol{v}}_t = \alpha \widehat{\boldsymbol{v}}_t + (1-\alpha)\widehat{\boldsymbol{v}}_{t-1}, \tag{3.42}$$

where α, $(0 < \alpha < 1)$ *is called the smoothing parameter.*

6. ***Iterate:*** *If the stopping criterion is met, stop; otherwise, set* $t = t+1$ *and return to Step 2.*

As a stopping criterion one can use, for example, the following: if for some $t \geq d$, say $d = 5$,

$$\widehat{b}_{t-1,(N)} = \widehat{b}_{t,(N)} = \cdots = \widehat{b}_{t-d,(N)} \tag{3.43}$$

then stop.

Remark 3.4

Note that as a modification of Algorithm 3.1 one can adopt the one where Step 3 is eliminated, that is, all samples are involved while updating $\widehat{\boldsymbol{v}}_t$, according to (3.41). The reason is that for large negative λ there is not too much difference for $\widehat{\boldsymbol{v}}_t$, whether we use the elite sample (as in Step 3) or the entire one (without Step 3).

3.5.2 Constrained Combinatorial Optimization: The Penalty Function Approach

Consider the following constrained combinatorial optimization problem (with inequality constraints):

$$\begin{aligned} \max_{\boldsymbol{x}} \ & \sum_{k=1}^{n} c_k x_k \\ \text{s.t.} \ & \sum_{k=1}^{n} a_{ik} x_k \geq b_i, \quad i = 1, \ldots, m \\ & x_k \in \{0, 1\} \quad \text{for all } k = 1, \ldots, n. \end{aligned} \tag{3.44}$$

Assume in addition that the vector $\boldsymbol{x}$ is binary and all components b_i and a_{ik} are positive numbers. Using the penalty method approach we can reduce the original constraint problem (3.44) to the following unconstrained one

$$\max_{\boldsymbol{x}} \left\{ S(\boldsymbol{x}) = \sum_{k=1}^{n} c_k x_k + M(\boldsymbol{x}, \beta) \right\}, \tag{3.45}$$

where the penalty function is defined as

$$M(\boldsymbol{x}, \beta) = \beta \sum_{i=1}^{m} I_{\{\sum_{k=1}^{n} a_{ik} x_k < b_i\}}. \tag{3.46}$$

Here,

$$\beta = a + \sum_{k=1}^{n} c_k, \tag{3.47}$$

where a is a positive number. If not stated otherwise, we assume that $a = 1$. Note that $M(\boldsymbol{x}, \beta)$ is an increasing function of the number of the constraints that $\boldsymbol{x}$ satisfies and the penalty parameter β is chosen such that $S(\boldsymbol{x})$ is negative when and only when at least one of the constraints is not satisfied. If $\boldsymbol{x}$ satisfies all the constraints in (3.44), then $M(\boldsymbol{x}, \beta) = 0$ and the $S(\boldsymbol{x})$ value is equal to the value of the original performance function in (3.44). Clearly, the optimization program (3.44) can again be associated with the rare-event probability estimation problem, where $\boldsymbol{X}$ is a vector of iid $\mathsf{Ber}(1/2)$ components.

We found that our numerical results with MinxEnt Algorithm 3.1 for different unconstrained and constrained combinatorial optimization problems show that it is comparable with its counterpart CE in both accuracy and the CPU time.

Chapter 4

Splitting Method for Counting and Optimization

4.1 BACKGROUND

This chapter deals with the *splitting* method for counting, combinatorial optimization, and rare-event estimation. Before turning to the splitting method we present some background on counting using randomized (or Monte Carlo) algorithms.

To date, very little is known about how to construct efficient algorithms for hard counting problems. This means that exact solutions to these problems cannot be obtained in polynomial time, and therefore our work focuses on approximation algorithms, and, in particular, approximation algorithms based on randomization. The basic procedure for counting is outlined below.

1. Formulate the counting problem as estimating the cardinality $|\mathcal{X}^*|$ of some set $\mathcal{X}^*$, such as the one in (1.1).
2. Find a sequence of decreasing sets $\mathcal{X} = \mathcal{X}_0, \mathcal{X}_1, \ldots, \mathcal{X}_m$ such that

$$\mathcal{X}_0 \supset \mathcal{X}_1 \supset \cdots \supset \mathcal{X}_m = \mathcal{X}^*, \tag{4.1}$$

   and $|\mathcal{X}| = |\mathcal{X}_0|$ is known.
3. Write $|\mathcal{X}^*| = |\mathcal{X}_m|$ as

$$|\mathcal{X}^*| = |\mathcal{X}_0| \prod_{t=1}^{m} \frac{|\mathcal{X}_t|}{|\mathcal{X}_{t-1}|} = \ell |\mathcal{X}_0|, \tag{4.2}$$

   where

$$\ell = \prod_{t=1}^{m} \frac{|\mathcal{X}_t|}{|\mathcal{X}_{t-1}|}. \tag{4.3}$$

Fast Sequential Monte Carlo Methods for Counting and Optimization, First Edition.
Reuven Y. Rubinstein, Ad Ridder, and Radislav Vaisman.

Note that ℓ is typically very small, like $\ell = 10^{-100}$, while each ratio

$$c_t = \frac{|\mathcal{X}_t|}{|\mathcal{X}_{t-1}|} \tag{4.4}$$

should be not too small, like $c_t = 10^{-2}$ or bigger. Clearly, estimating ℓ directly while sampling in $|\mathcal{X}_0|$ is meaningless, but estimating each c_t separately seems to be a good alternative.

4. Develop an efficient estimator $\widehat{c}_t$ for each c_t and estimate $|\mathcal{X}^*|$ by

$$\widehat{|\mathcal{X}^*|} = \widehat{\ell}|\mathcal{X}_0| = |\mathcal{X}_0| \prod_{t=1}^{m} \widehat{c}_t, \tag{4.5}$$

where $\widehat{\ell} = \prod_{t=1}^{m} \widehat{c}_t$.

Using (4.1)–(4.5), one can design a sampling plan according to which the "difficult" counting problem defined on the set $\mathcal{X}^*$ is decomposed into a number of "easy" ones associated with a sequence of related sets $\mathcal{X}_0, \mathcal{X}_1, \ldots, \mathcal{X}_m$ and such that $\mathcal{X}_m = \mathcal{X}^*$. Typically, the Monte Carlo algorithms based on (4.1)–(4.5) explore the connection between counting and sampling and in particular the reduction from approximate counting of a discrete set to approximate sampling of the elements of this set [108].

To deliver a meaningful estimator of $|\mathcal{X}^*|$, we must solve the following two major problems:

i. Construct the sequence $\mathcal{X}_0 \supset \mathcal{X}_1 \supset \cdots \supset \mathcal{X}_m = \mathcal{X}^*$ such that each c_t is not a rare-event probability.

ii. Obtain a low variance unbiased estimator $\widehat{c}_t$ of each $c_t = |\mathcal{X}_t|/|\mathcal{X}_{t-1}|$.

Hence, once both tasks (i) and (ii) are resolved, one can obtain an efficient estimator for $\widehat{|\mathcal{X}^*|}$ given in (4.5).

Task (i) might be typically simply resolved in many cases. Let us consider a concrete example:

- Suppose that each variable of the set (1.1) is bounded, say $L_j \leq x_j \leq \mathrm{U}_j$ $(L_j, \mathrm{U}_j \in \mathbb{Z})$, then we may take $\mathcal{X}_0 = \prod_{j=1}^{n}[L_j, \mathrm{U}_j]$, with $|\mathcal{X}_0| = \prod_{j=1}^{n}(\mathrm{U}_j - L_j + 1)$.
- Consider the function $S(\boldsymbol{x})$ to be the number of constraints of the integer program (1.1); hence, $S : \mathcal{X}_0 \to \{0, 1, \ldots, m\}$.
- Define finally the desired subregions $\mathcal{X}_t$ by

$$\mathcal{X}_t = \{\boldsymbol{x} \in \mathcal{X}_0 \ : \ S(\boldsymbol{x}) \geq t\}, \quad t = 0, 1, \ldots, m.$$

Task (ii) is quite complicated. It is associated with uniform sampling separately at each subregion $\mathcal{X}_t$ and will be addressed in the subsequent text.

The rest of this chapter is organized as follows. Section 4.2 presents a quick look at the splitting method. We show that, similarly to the conventional classic randomized algorithms (see [88]), it uses a sequential sampling plan to decompose a "difficult" problem into a sequence of "easy" ones. Sections 4.3 and 4.4 present two different splitting algorithms for counting; the first based on a fixed-level set and the second on an adaptive level set. Both algorithms employ a combination of a Markov chain Monte Carlo algorithm, such as the Gibbs sampler, with a specially designed cloning mechanism. The latter runs in parallel multiple Markov chains by making sure that all of them run in steady-state at each iteration. Section 4.5 shows how, using the splitting method, one can generate a sequence of points $\boldsymbol{X}_1, \ldots, \boldsymbol{X}_N$ uniformly distributed on a discrete set $\mathcal{X}^*$. By uniformity we mean that the sample $\boldsymbol{X}_1, \ldots, \boldsymbol{X}_N$ on $\mathcal{X}^*$ passes the standard Chi-squared test [91]. We support numerically the uniformity of generated samples based on the Chi-squared test.

In Section 4.6 we show that the splitting method is suitable for solving combinatorial optimization problems, such as the maximal cut or traveling salesman problems, and thus can be considered as an alternative to the standard cross-entropy and MinxEnt methods. Sections 4.7.1 and 4.7.2 deal with two enhancements of the adaptive splitting method for counting. The first is called the direct splitting estimator and is based on the direct counting of $|\mathcal{X}^*|$, while the second is called the capture-recapture estimator of $|\mathcal{X}^*|$. It has its origin in the well-known capture-recapture method in statistics [114]. Both are used to obtain low-variance estimators for counting on complex sets as compared to the adaptive splitting algorithm. Section 4.8 shows how the splitting algorithm can be efficiently used for estimating the reliability of complex static networks. Finally, Section 4.9 presents supportive numerical results for counting, rare events, and optimization.

4.2 QUICK GLANCE AT THE SPLITTING METHOD

In this section we will take a quick look at our sampling mechanism using splitting. Consider the counting problem (4.2) with T subsets, that is,

$$|\mathcal{X}^*| = |\mathcal{X}| \prod_{t=1}^{T} \frac{|\mathcal{X}_t|}{|\mathcal{X}_{t-1}|} = \ell |\mathcal{X}|,$$

where $\ell = \prod_{t=1}^{T} |\mathcal{X}_t| / |\mathcal{X}_{t-1}|$. We assume that the subsets $\mathcal{X}_t$ are associated with levels m_t and can be written as

$$\mathcal{X}_t = \{\boldsymbol{x} \in \mathcal{X} : S(\boldsymbol{x}) \geq m_t\} \quad t = 1, \ldots, T,$$

where $S : \mathcal{X} \rightarrow \mathbb{R}$ is the sample performance function and $m_1 \leq m_2 \leq \cdots \leq m_T = m$. We set for convenience $m_0 = -\infty$. The other levels are either fixed in advance, or their values are determined adaptively by the splitting algorithm during the course of simulation. Observe that ℓ in (4.3) can be interpreted as

$$\ell = \mathbb{E}_f \left[I_{\{S(\boldsymbol{X}) \geq m\}} \right], \tag{4.6}$$

where $\boldsymbol{X} \sim f(\boldsymbol{x})$, the uniform distribution on $\mathcal{X}$. We denote this by $f = \mathsf{U}(\mathcal{X})$. Furthermore, we shall require a uniform pdf on each subset $\mathcal{X}_t$, which is denoted by $g_t^*(\cdot) = g^*(\cdot, m_t) = \mathsf{U}(\mathcal{X}_t)$. Clearly,

$$g_t^*(\boldsymbol{x}) = g^*(\boldsymbol{x}, m_t) = \ell(m_t)^{-1} f(\boldsymbol{x}) I_{\{S(\boldsymbol{x}) \geq m_t\}}, \tag{4.7}$$

where $\ell(m_t)^{-1}$ is the normalization constant. Generating points uniformly distributed on $\mathcal{X}_t$ using the splitting method will be addressed in Section 4.5. With these notations, we can consider c_t of (4.4) to be a conditional expectation defined as

$$c_t = \mathbb{E}_{g_{t-1}^*}\left[I_{\{S(\boldsymbol{X}) \geq m_t\}}\right] = \mathbb{E}_f\left[I_{\{S(\boldsymbol{X}) \geq m_t\}} | S(\boldsymbol{X}) \geq m_{t-1}\right] \quad t = 1, \ldots, T. \tag{4.8}$$

Suppose that sampling from $g_t^* = \mathsf{U}(\mathcal{X}_t)$ becomes available, and that we generated $N_t^{(e)}$ samples in $\mathcal{X}_t$, then the final estimators of ℓ and $|\mathcal{X}^*|$ can be written as

$$\widehat{\ell} = \prod_{t=1}^{T} \widehat{c}_t = \frac{1}{N^T} \prod_{t=1}^{T} N_t^{(e)}, \tag{4.9}$$

and

$$\widehat{|\mathcal{X}^*|} = |\mathcal{X}| \widehat{\ell} = \frac{|\mathcal{X}|}{N^T} \prod_{t=1}^{T} N_t^{(e)}, \tag{4.10}$$

respectively. Here we used

$$\widehat{c}_t = \frac{1}{N} \sum_{i=1}^{N} I_{\{S(\boldsymbol{X}_i) \geq m_t\}} = \frac{N_t^{(e)}}{N} \tag{4.11}$$

as estimators of their true unknown conditional expectations. Furthermore, $N_t^{(e)} = \sum_{i=1}^{N} I_{\{S(\boldsymbol{X}_i) \geq m_t\}}$, with $\boldsymbol{X}_i \sim g_{t-1}^*$ and $g_0^* = f$. We call $N_t^{(e)}$ the *elite sample size*, associated with the *elite samples*.

To provide more insight into the splitting method, consider a toy example of the integer problem (1.1), pictured in Figure 4.1. The solution set is a six-sided polytope in the two-dimensional plane, obtained by $m = 6$ inequality constraints. Its boundary is shown by the bold lines in the figure. The bounding set is the unit square $\mathcal{X}_0 = [0, 1] \times [0, 1]$. We generate $N = 5$ points uniformly in $\mathcal{X}_0$ (see the dots in the figure). The corresponding five performance values $S_i = S(\boldsymbol{X}_i)$, $i = 1, \ldots, 5$ are

$$S_1 = 5, S_2 = 4, S_3 = 3, S_4 = 3, S_5 = 2.$$

As for elite samples, we choose the two largest values corresponding to $S_1 = 5$ and $S_2 = 4$. We thus have $N_1^{(e)} = 2$, and $m_1 = 4$. Consider the first elite sample $\boldsymbol{X}_1$ with $S_1 = 5$. Its associated subregion of points satisfying exactly the same five constraints is given by the shaded area in Figure 4.2.

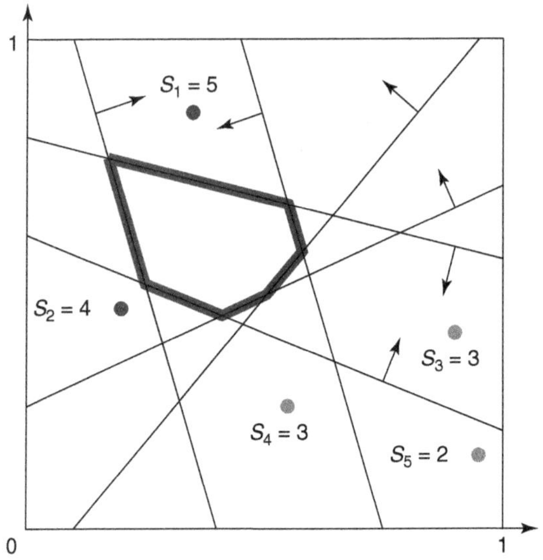

Figure 4.1 Iteration 1: five uniform samples on $\mathcal{X}_0$.

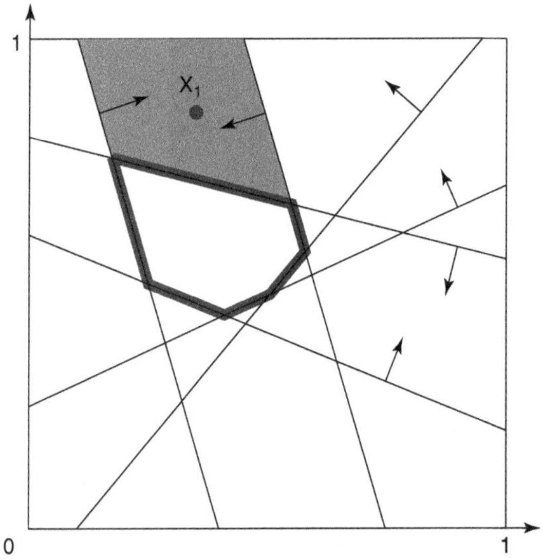

Figure 4.2 The subregion corresponding to the elite point $\boldsymbol{X}_1$ with $S = 5$.

Similarly, Figure 4.3 shows the elite point $\boldsymbol{X}_2$ (with $S_2 = 4$) and its associated subregion of points satisfying the same four constraints. However, notice that there are more subregions of points satisfying at least four constraints. The union of all these subregions is the set $\mathcal{X}_1$; that is,

$$\mathcal{X}_1 = \{\boldsymbol{x} : S(\boldsymbol{x}) \geq 4 = m_1\},$$

as shown in Figure 4.4.

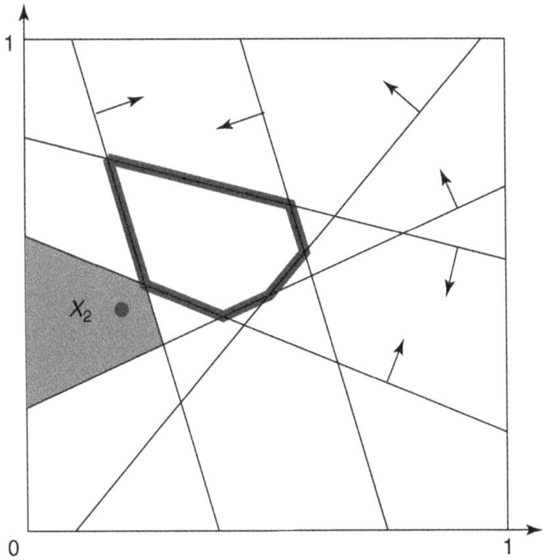

Figure 4.3 The subregion corresponding to the elite point $\boldsymbol{X}_2$ with $S = 4$.

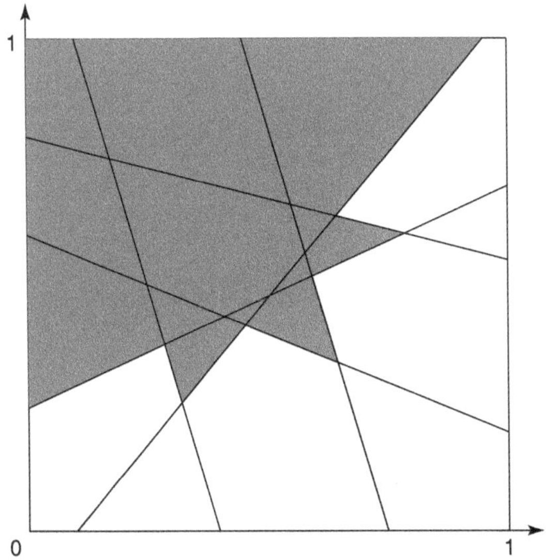

Figure 4.4 The subregion $\mathcal{X}_1$ containing all points with $S \geq 4$.

Starting from the two elite points $\boldsymbol{X}_1, \boldsymbol{X}_2 \in \mathcal{X}_1$, we apply a Markov chain Monte Carlo algorithm that generates points in $\mathcal{X}_1$ such that these points are uniformly distributed. Figure 4.5 shows $N = 5$ resulting points. The five performance values are

$$S_1 = 5, S_2 = 5, S_3 = 4, S_4 = 4, S_5 = 4.$$

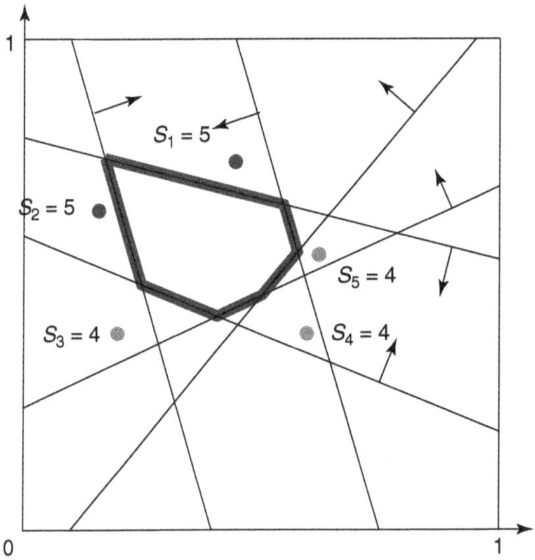

Figure 4.5 Iteration 2: five uniform points on $\mathcal{X}_1$.

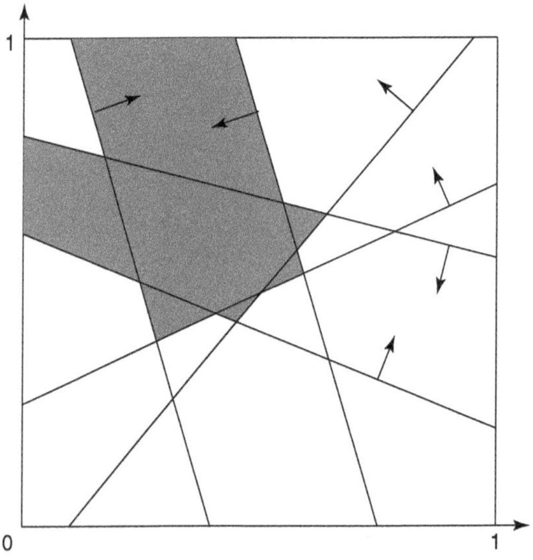

Figure 4.6 The subregion $\mathcal{X}_2$ containing all points with $S \geq 5$.

Next, we repeat the same procedure as in the first iteration. As elite samples we choose the two largest points $S_1 = 5$, $S_2 = 5$. Thus, $N_2^{(e)} = 2$, and $m_2 = 5$. This gives us the set of all points satisfying at least five constraints,

$$\mathcal{X}_2 = \{\boldsymbol{x} : S(\boldsymbol{x}) \geq 5 = m_2\};$$

see Figure 4.6.

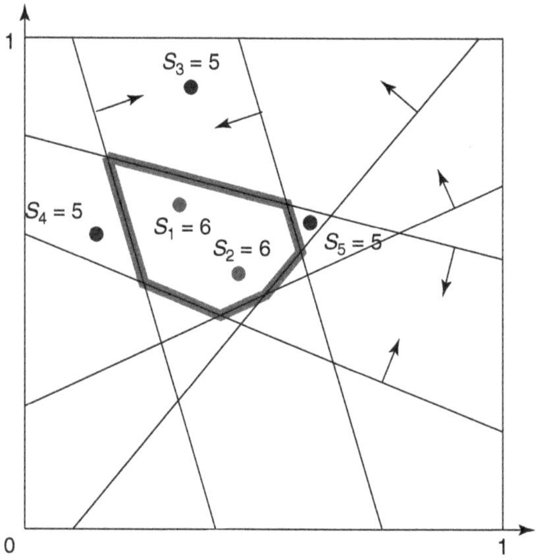

Figure 4.7 Iteration 3: five uniform points on $\mathcal{X}_2$.

We use a Markov chain Monte Carlo algorithm to generate from the two elite points five new points in $\mathcal{X}_2$, uniformly distributed; see Figure 4.7. Note that two of the new points hit the desired polytope. At this stage, we stop the iterations and conclude that the algorithm has converged. The purpose of Figures 4.1–4.7 is to demonstrate that the splitting method can be viewed as a global search in the sense that, at each iteration, the generated random points $\boldsymbol{X}_i$, $i = 1, \ldots, 5$ are randomly distributed inside their corresponding subregion. The main issue remaining in the forthcoming sections is to provide exact algorithms that implement the splitting method.

Before doing so, we show how to cast the problem of counting the number of what we call multiple events into the framework (4.6)–(4.8). As we shall see below, counting the number of feasible points $|\mathcal{X}^*|$ of the set (1.1) follows as a particular case of it.

EXAMPLE 4.1 ***Counting Multiple Events***

Consider counting the cardinality of the set

$$\mathcal{X}^* = \{\boldsymbol{x} \in \mathbb{R}^n : S_i(\boldsymbol{x}) \geq b_i, \ i = 1, \ldots, m\}, \tag{4.12}$$

where $S_i : \mathbb{R}^n \to \mathbb{R}$, $i = 1, \ldots, m$ are arbitrary functions. In this case, we can associate with (4.12) the following multiple-event probability estimation problem:

$$\ell = \mathbb{P}_f\left(\bigcap_{i=1}^{m} \{S_i(\boldsymbol{X}) \geq b_i\}\right) = \mathbb{E}_f\left[\prod_{i=1}^{m} I_{\{S_i(\boldsymbol{X}) \geq b_i\}}\right]. \tag{4.13}$$

Note that (4.13) extends (4.6) in the sense that it involves an intersection of m events $\{S_i(\boldsymbol{X}) \geq b_i\}$, that is, multiple events rather than a single one $\{S(\boldsymbol{X}) \geq b\}$. Some of the constraints may be equality constraints, $\{S_i(\boldsymbol{X}) = b_i\}$.

As before, we assume that each individual indicator event in (4.13) is not rare, that is, each probability of the type $\mathbb{P}_f(S_i(\boldsymbol{X}) \geq b_i)$ is not a rare-event probability, say $\mathbb{P}_f(S_i(\boldsymbol{X}) \geq b_i) \geq 10^{-3}$, but the event intersection forms a rare-event probability ℓ.

It is clear that in this case ℓ in (4.13) can be rewritten as

$$\ell = \mathbb{E}_f \left[I_{\{\sum_{i=1}^m C_i(\boldsymbol{X})=m\}} \right], \tag{4.14}$$

where

$$C_i(\boldsymbol{X}) = I_{\{S_i(\boldsymbol{X})\geq b_i\}}, \quad i = 1, \ldots, m. \tag{4.15}$$

For the cardinality of the set $\mathcal{X}^*$ one applies (4.10), which means that one only needs to estimate the rare-event probability ℓ in (4.14) according to (4.9). □

4.3 SPLITTING ALGORITHM WITH FIXED LEVELS

In this section we assume that the levels m_t, $t = 1, \ldots, T$ are fixed in advance, with $m_T = m$. This can be done using a pilot run with the splitting method as in [73]. Consider the problem of estimating the conditional probability

$$c_{t+1} = \mathbb{P}(S(\boldsymbol{X}) \geq m_{t+1} | S(\boldsymbol{X}) \geq m_t) = \mathbb{P}(\boldsymbol{X} \in \mathcal{X}_{t+1} | \boldsymbol{X} \in \mathcal{X}_t).$$

Assume further that we have generated a point $\boldsymbol{X} \in \mathcal{X}_t$ uniformly distributed on $\mathcal{X}_t$, that is, $\boldsymbol{X} \sim \mathsf{U}(\mathcal{X}_t)$. Recall from our toy example in the previous section that, given this information of the current point, we should be able to generate a new point in $\mathcal{X}_t$, which is again uniformly distributed. Typically, this might be obtained by applying a Markov chain Monte Carlo technique. For now we just assume that we dispose of a mapping $\Phi_t : \mathcal{X}_t \to \mathcal{X}_t$, which preserves uniformity, that is,

$$\boldsymbol{X} \sim \mathsf{U}(\mathcal{X}_t) \quad \Rightarrow \quad \Phi_t(\boldsymbol{X}) \sim \mathsf{U}(\mathcal{X}_t). \tag{4.16}$$

We call this mapping the *uniform mutation mapping*. Based on that we can derive an unbiased estimator of c_{t+1} by using the following procedure:

1. Generate a sample $\boldsymbol{X}_1, \ldots, \boldsymbol{X}_{N_t}$ of points uniformly distributed on $\mathcal{X}_t$.
2. Apply the uniform mutation mapping to each point, and determine the cardinality $N_{t+1}^{(e)}$ of the set of points $\{\Phi_t(\boldsymbol{X}_i) \in \mathcal{X}_{t+1}\}$. This set is called the *elite set*.
3. Deliver

$$\widehat{c}_{t+1} = \frac{N_{t+1}^{(e)}}{N_t} = \frac{1}{N_t}\sum_{i=1}^{N_t} I_{\{\Phi_t(X_i)\in\mathcal{X}_{t+1}\}}$$

as an unbiased estimator of c_{t+1}.

The following algorithm applies this procedure iteratively. Moreover, in each iteration all $N_{t+1}^{(e)}$ elite points are reproduced (cloned) η_{t+1} times to form a new sample of $N_{t+1} = \eta_{t+1} N_{t+1}^{(e)}$ points. The parameter η_{t+1} is called the *splitting parameter*.

Algorithm 4.1 ***Fixed-Level Splitting Algorithm for Counting***

- *Input: the counting problem* (1.1)*; the fixed levels* $m_t,\ t = 0, 1, \ldots, T;\ m_0 = -\infty;\ m_T = m$*; the splitting parameters* $\eta_t,\ t = 1, \ldots, T-1$*; the initial sample size* N_0.
- *Output: estimators* (4.9) *and* (4.10).

1. ***Initialize:*** *Set a counter* $t = 0$*. Generate a sample* $[\boldsymbol{X}]_0 = \{\boldsymbol{X}_1, \ldots, \boldsymbol{X}_{N_0}\}$ *uniformly on* $\mathcal{X}_0$.
2. ***Select:*** *Determine the elite sample, that is, the points for which* $S(\boldsymbol{X}_i) \geq m_{t+1}$*. Suppose that* $N_{t+1}^{(e)}$ *is the size of this subset, and denote these points by* $\boldsymbol{X}_j^{(e)}$*; thus,*

$$[\boldsymbol{X}^{(e)}]_{t+1} = \{\boldsymbol{X}_1^{(e)}, \ldots, \boldsymbol{X}_{N_{t+1}^{(e)}}^{(e)}\}.$$

   Note that each elite $\boldsymbol{X}^{(e)} \sim \mathsf{U}(\mathcal{X}_{t+1})$.
3. ***Estimate*** c_{t+1}***:*** *Take*

$$\widehat{c}_{t+1} = \frac{1}{N_t}\sum_{i=1}^{N_t} I_{\{S(X_i)\geq m_{t+1}\}} = \frac{N_{t+1}^{(e)}}{N_t}$$

   as an estimator of c_{t+1}.
4. ***Stop:*** *If* $t + 1 = T$ $(m_{t+1} = m)$*, deliver the estimators* (4.9) *and* (4.10)*. Else, continue with the next step.*
5. ***Splitting:*** *Reproduce (clone)* η_{t+1} *times each sample point* $\boldsymbol{X}^{(e)}$ *of the elite sample. Set* $N_{t+1} = N_{t+1}^{(e)}\eta_{t+1}$.
6. ***Mutation:*** *To each of the cloned points apply the uniform mutation mapping* Φ_{t+1}*. Denote the new sample by* $[\boldsymbol{X}]_{t+1} = \{\boldsymbol{X}_1, \ldots, \boldsymbol{X}_{N_{t+1}}\}$*. Note that each point in the sample is distributed uniformly on* $\mathcal{X}_{t+1}$.
7. ***Iterate:*** *Increase the counter* $t = t + 1$*, and repeat from step 2.*

A key ingredient in the Algorithm 4.1 is the uniform mutation mapping Φ_t. Typically, this is obtained by a Markov chain kernel $K_t(\boldsymbol{x}, A)$ on $\mathcal{X}_t$ with the uniform distribution as its invariant distribution. This means that we assume:

i. There is some aperiodic irreducible Markov chain $\{Y_n, n = 0, 1, \dots\}$ on $\mathcal{X}_t$ with transition probabilities

$$K_t(\boldsymbol{x}, A) = \mathbb{P}(Y_{n+1} \in A | Y_n = \boldsymbol{x})$$

for measurable subsets $A \subset \mathcal{X}_t$.

ii. The invariant probability distribution $\pi(\cdot)$ exists; that is,

$$\pi(A) = \int_{\mathcal{X}_t} K_t(\boldsymbol{x}, A)\pi(\boldsymbol{x})\, d\boldsymbol{x} \quad (A \subset \mathcal{X}_t).$$

iii. $\pi(\cdot)$ is the uniform distribution on $\mathcal{X}_t$.

Under these assumptions, we can generate from any point $\boldsymbol{x} \in \mathcal{X}_t$ a path of points $Y_0 = \boldsymbol{x}, Y_1, Y_2, \dots$ in $\mathcal{X}_t$, by repeatedly applying the Markov kernel $K_t(\cdot, \cdot)$. If the initial point is random uniformly distributed, then each consecutive point of the path preserves this property (assumption (iii)). Otherwise, we use the property that the limiting distribution of Markov chain is its invariant distribution. In implementations we often resort to the Gibbs sampler for constructing the Markov kernel and generating an associated path of points (see Appendix 4.10 for details). All these considerations are summarized in just mentioning "apply the uniform mutation mapping."

We next show that the estimator $\widehat{\ell}$ is unbiased. Indeed taking into account that

$$\ell = \prod_{t=1}^{T} c_t\,;$$

$$\widehat{\ell} = \prod_{t=1}^{T} \widehat{c}_t = \prod_{t=0}^{T-1} \widehat{c}_{t+1} = \prod_{t=0}^{T-1} \frac{N_{t+1}^{(e)}}{N_t} = \frac{N_1^{(e)}}{N_0} \prod_{t=1}^{T-1} \frac{N_{t+1}^{(e)}}{\eta_t N_t^{(e)}}$$

$$= \frac{1}{N_0 \prod_{t=1}^{T-1} \eta_t} N_T^{(e)}\,;$$

$$\mathbb{E}[N_t^{(e)} | N_{t-1}^{(e)}] = \mathbb{E}[N_t^{(e)} | N_{t-1} = \eta_{t-1} N_{t-1}^{(e)}] = c_t N_{t-1} = c_t \eta_{t-1} N_{t-1}^{(e)};$$

$$\mathbb{E}[N_1^{(e)}] = c_1 N_0,$$

we obtain from the last two equalities that

$$\mathbb{E}\left[N_T^{(e)}\right] = \mathbb{E}\left[\mathbb{E}\left[N_T^{(e)} | N_{T-1}^{(e)}\right]\right] = c_T \eta_{T-1} \mathbb{E}\left[N_{T-1}^{(e)}\right]$$

$$= \cdots = \left(\prod_{t=2}^{T} c_t \eta_{t-1}\right) \mathbb{E}[N_1^{(e)}] = \left(\prod_{t=2}^{T} c_t \eta_{t-1}\right) c_1 N_0,$$

from which $\mathbb{E}[\widehat{\ell}] = \ell$ readily follows. We next calculate the variance of $\widehat{\ell}$. In order to accomplish this, we do the following:

1. Find for each point $\boldsymbol{X}^{(e)} \in [\boldsymbol{X}^{(e)}]_1$ in the first elite set its corresponding number Z of offspring in the final subset $\mathcal{X}_T = \mathcal{X}^*$.
2. Denote by Z_i the number of offspring generated from the i-th elite point. Then clearly,

$$N_T^{(e)} = \sum_{i=1}^{N_1^{(e)}} Z_i,$$

   and $Z_1, Z_2, \ldots$ are iid.
3. Calculate $\mathbb{V}\mathrm{ar}[\widehat{\ell}]$ as

$$\mathbb{V}\mathrm{ar}[\widehat{\ell}] = \frac{1}{N_0^2 \prod_{t=1}^{T-1} \eta_t^2} \mathbb{V}\mathrm{ar}[N_T^{(e)}], \tag{4.17}$$

   where

$$\mathbb{V}\mathrm{ar}[N_T^{(e)}] = N_0 c_1 \mathbb{V}\mathrm{ar}[Z] + N_0 c_1 (1 - c_1)(\mathbb{E}[Z])^2$$

   and

$$\mathbb{E}[Z] = \frac{\mathbb{E}[N_T^{(e)}]}{N_0 c_1}.$$

Thus, to estimate the variance of $\widehat{\ell}$ it suffices to estimate $\mathbb{V}\mathrm{ar}[Z]$, which in turn can be calculated as

$$S^2 = \frac{1}{N_1^{(e)} - 1} \sum_{i=1}^{N_1^{(e)}} (Z_i - \overline{Z})^2,$$

where

$$\overline{Z} = \frac{1}{N_1^{(e)}} \sum_{i=1}^{N_1^{(e)}} Z_i,$$

and $Z_1, Z_2, \ldots$ are iid samples.

It is challenging to find good splitting parameters η_t in the fixed-level splitting version. When these parameters are not chosen correctly, the number of samples will either die out or explode, leading in both cases to large variances of $\widehat{\ell}$. The minimal variance choice is $\eta_t c_t = 1$, see [48] and [55]. Thus, given the levels m_t, $t = 1, 2, \ldots$, we can execute pilot runs to find rough estimates of the c_t's, denoted by ρ_t, $t = 1, \ldots, T$. At iteration $t - 1$ ($t = 1, 2, \ldots$) of the algorithm we set the splitting parameter of the i-th elite point to be a random variable defined as

$$\eta_t(i) = \lfloor 1/\rho_t \rfloor + B_i, \quad i = 1, \ldots, N_t^{(e)};$$

where the variable B_i is a Bernoulli random variable with success probability $(1/\rho_t) - \lfloor 1/\rho_t \rfloor$, and $\lfloor \cdot \rfloor$ means rounding down to the nearest integer. Thus, the splitting parameters satisfy $\mathbb{E}[\eta_t] = 1/\rho_t \approx 1/c_t$. Note that the sample size in the next iteration is also a random variable defined as

$$N_t = \sum_{i=1}^{N_t^{(e)}} \eta_t(i).$$

In the following section we propose an adaptive splitting method considering two issues that can cause difficulties in the fixed-level approach:

- An objection against cloning is that it introduces correlation. An alternative approach is to sample η_t "new" points in the current subset by running a Markov chain using a Markov chain Monte Carlo method.
- Finding "good" locations of the levels m_t, $t = 1, 2, \ldots$ is crucial but difficult. We propose to select the levels adaptively.

4.4 ADAPTIVE SPLITTING ALGORITHM

In both counting and optimization problems the proposed adaptive splitting algorithms generate sequences of pairs

$$(m_1, g^*(\boldsymbol{x}, m_0)),\ (m_2, g^*(\boldsymbol{x}, m_1)), \ldots, (m_T, g^*(\boldsymbol{x}, m_{T-1})), \tag{4.18}$$

where, as before, $g^*(\boldsymbol{x}, m_0) = f(\boldsymbol{x})$. This is in contrast to the cross-entropy method [107], where one generates a sequence of pairs

$$(m_0, \boldsymbol{v}_0),\ (m_1, \boldsymbol{v}_1), \ldots, (m_T, \boldsymbol{v}_T). \tag{4.19}$$

Here $\boldsymbol{v}_t$, $t = 1, \ldots, T$ is a sequence of parameters in the parametric family of distributions $f(\boldsymbol{x}, \boldsymbol{v}_t)$. The crucial difference is, of course, that in the splitting method, $g^*(\boldsymbol{x}, m_t) = g_t^*(\boldsymbol{x})$, $t = 1, \ldots, T$ is a sequence of (uniform) nonparametric distributions rather than of parametric ones $f(\boldsymbol{x}, \boldsymbol{v}_t)$, $t = 1, \ldots, T$. Otherwise the cross-entropy and the splitting algorithms are very similar, apart from the fact that cross-entropy has the desirable property that the samples are independent, whereas in splitting they are not. The advantage of the splitting method is that it permits approximate sampling from the uniform pdf $g^*(\boldsymbol{x}, m_t) = g_t^*$ and permits updating the parameters c_t and m_t adaptively. It is shown in [104] that for rare-event estimation and counting the splitting method typically outperforms its cross-entropy counterpart, while for combinatorial optimization they perform similarly. The main reason is that for counting the sampling from a sequence of nonparametric probability density function g_t^* is more beneficial (more informative) than sampling from a sequence of a parametric family like $f(\boldsymbol{x}, \boldsymbol{v}_t)$.

Below we present the main steps of the adaptive splitting algorithm. However, first, we describe the main ideas for which we refer to [13] and [40], where the adaptive (m_t)-thresholds satisfy the requirements $c_t = |\mathcal{X}_t|/|\mathcal{X}_{t-1}| \approx \rho_t$. Here, as before, the parameters $\rho_t \in (0, 1)$, called the *rarity* parameters, are fixed in advance and they must be not too small, say $\rho_t \geq 0.01$.

We start by generating a sample of N points uniformly on $\mathcal{X}_0$, denoted by $[\boldsymbol{X}]_0 = \{\boldsymbol{X}_1, \ldots, \boldsymbol{X}_N\}$. Consider the sample set at the t-th iteration: $[\boldsymbol{X}]_t = \{\boldsymbol{X}_1, \ldots, \boldsymbol{X}_N\}$ of N random points in $\mathcal{X}_t$. All these sampled points are *uniformly distributed* on $\mathcal{X}_t$. Let m_{t+1} be the $(1 - \rho_{t+1})$-th quantile of the ordered statistics values of the values $S(\boldsymbol{X}_1), \ldots, S(\boldsymbol{X}_N)$. The *elite set* $[\boldsymbol{X}^{(e)}]_{t+1} \subset [\boldsymbol{X}]_t$ consists of those points of the sample set for which $S(\boldsymbol{X}_i) \geq m_{t+1}$. Let $N_{t+1}^{(e)}$ be the size of the elite set. If all values $S(\boldsymbol{X}_i)$ would be distinct, it would follow that the number of elites $N_{t+1}^{(e)} = \lceil N\rho_{t+1} \rceil$, where $\lceil \cdot \rceil$ denotes rounding up to the nearest integer. However, when we deal with a discrete finite space, typically we will find many duplicate sampled points with $S(\boldsymbol{X}_i) \geq m_{t+1}$. All these are included in the elite set. Level m_{t+1} is used to define the next level subset $\mathcal{X}_{t+1} = \{\boldsymbol{x} \in \mathcal{X} : S(\boldsymbol{x}) \geq m_{t+1}\}$. Finally, note that it easily follows from (4.7) that the elite points are distributed uniformly on this level subset.

Regarding the elite sample $[\boldsymbol{X}^{(e)}]_{t+1}$, we do three things. First, we use its relative size $N_{t+1}^{(e)}/N$ as an estimate of the conditional probability $c_{t+1} = \mathbb{P}(S(\boldsymbol{X}) \geq m_{t+1} \mid \boldsymbol{X} \in \mathcal{X}_t)$. Next, we delete duplicates. We call this *screening*, and it might be modeled formally by introducing the identity map $\sigma : \mathcal{X} \to \mathcal{X}$; that is, $\sigma(\boldsymbol{X}) = \boldsymbol{X}$ for all points $\boldsymbol{X} \in \mathcal{X}$. Apply the identity map σ to the elite set, resulting in a set of $N_{t+1}^{(s)}$ distinct points (called the screened elites), that we denote by

$$[\boldsymbol{X}^{(s)}]_{t+1} = \sigma([\boldsymbol{X}^{(e)}]_{t+1}) = \{\boldsymbol{X}_1^{(s)}, \ldots, \boldsymbol{X}_{N_{t+1}^{(s)}}^{(s)}\} \subset \mathcal{X}_{t+1}. \tag{4.20}$$

Thirdly, each screened elite is the starting point of η_{t+1} trajectories in $\mathcal{X}_{t+1}$ generated by a *Markov chain simulation* using a transition probability matrix with $g_{t+1}^* = \mathsf{U}(\mathcal{X}_{t+1})$ as its stationary distribution. Because the starting point is uniformly distributed, all consecutive points on the sample paths are uniformly distributed on $\mathcal{X}_{t+1}$. Therefore, we may use all these points in the next iteration. If the splitting parameter $\eta_{t+1} = 1$, all trajectories are independent.

Thus, we simulate $N_{t+1}^{(s)}\eta_{t+1}$ trajectories, each trajectory for $b_{t+1} = \lfloor N/(N_{t+1}^{(s)}\eta_{t+1}) \rfloor$ steps. This produces a total of $N_{t+1}^{(s)}\eta_{t+1}b_{t+1} \leq N$ uniform points in $\mathcal{X}_{t+1}$. To continue with the next iteration again with a sample set of size N, we choose randomly $N - N_{t+1}^{(s)}\eta_{t+1}b_{t+1}$ of the generated trajectories (without replacement) and apply one more Markov chain transition to the end points of these trajectories. Denote the new sample set by $[\boldsymbol{X}]_{t+1}$, and repeat the same procedure as above. Summarizing the above:

$$\mathcal{X}_t \supset [\boldsymbol{X}]_t \supset \underbrace{[\boldsymbol{X}^{(e)}]_{t+1} \supset [\boldsymbol{X}^{(s)}]_{t+1}}_{\subset \mathcal{X}_{t+1}}. \tag{4.21}$$

The algorithm iterates until it converges to $m_t = m$, say at iteration T, at which stage we stop and deliver the estimators (4.9) and (4.10) of ℓ and $|\mathcal{X}^*|$, respectively.

Figure 4.8 represents a typical dynamics of the zero-th iteration of the adaptive splitting algorithm below for counting. Initially $N = 50$ points were generated randomly in the two-dimensional space $\mathcal{X}_0$ (some are shown as black dots). Because

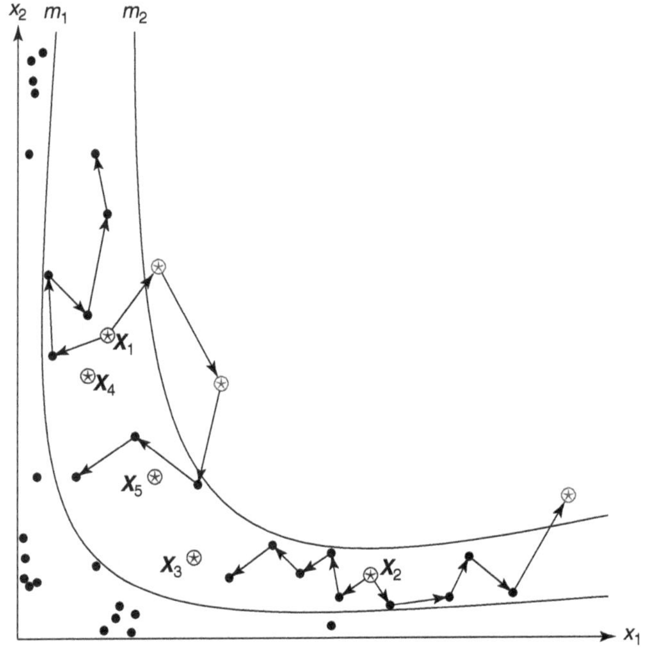

Figure 4.8 Iteration of the splitting algorithm.

$\rho_1 = 0.1$, the elite sample consists of $N_1^{(e)} = 5$ points (shown as the ⊛ points numbered $\boldsymbol{X}_1, \cdots, \boldsymbol{X}_5$) with values $S(\boldsymbol{x}) \geq m_1$. After renumbering, these five points are $\boldsymbol{X}_1, \cdots, \boldsymbol{X}_5$. From each of these elite points $\eta_1 = 2$ Markov chain paths of length $b_1 = 5$ were generated resulting in again $N = \eta_1 b_1 N_1^{(e)} = 50$ points in the space $\mathcal{X}_1 = \{\boldsymbol{x} \in \mathcal{X}_0 : S(\boldsymbol{x}) \geq m_1\}$. In the picture we see these paths from two points $\boldsymbol{X}_1$ and $\boldsymbol{X}_2$. Three points (the ⊛ in the upper region) are shown reaching the next-level m_2.

Algorithm 4.2 ***Adaptive Splitting Algorithm for Counting***

- *Input: The counting problem* (4.2)*; the rarity parameters* ρ_t*,* $t = 1, 2, \ldots$*; the splitting parameters* η_t*,* $t = 1, 2, \ldots$*; the sample size* N*.*
- *Output: Estimators* (4.9) *and* (4.10)*.*

1. ***Initialize:*** *Set a counter* $t = 0$*. Generate a sample* $[\boldsymbol{X}]_0 = \{\boldsymbol{X}_1, \ldots, \boldsymbol{X}_N\}$ *uniformly on* $\mathcal{X}_0$*.*
2. ***Select:*** *Compute level* $\widehat{m}_{t+1}$ *as the* $(1 - \rho_{t+1})$ *quantile of the ordered statistics values of* $S(\boldsymbol{X}_1), \ldots, S(\boldsymbol{X}_N)$*. Determine the elite sample, that is, the largest subset of* $[\boldsymbol{X}]_t$ *consisting of points for which* $S(\boldsymbol{X}_i) \geq \widehat{m}_{t+1}$*. Suppose that* $N_{t+1}^{(e)}$ *is the size of this subset, and denote its points by* $\boldsymbol{X}_j^{(e)}$*; thus,*

$$[\boldsymbol{X}^{(e)}]_{t+1} = \{\boldsymbol{X}_1^{(e)}, \ldots, \boldsymbol{X}_{N_{t+1}^{(e)}}^{(e)}\}.$$

Note that each elite point $X^{(e)} \sim g^*(x, \widehat{m}_{t+1}) = \mathsf{U}(\mathcal{X}_{t+1})$*, the uniform distribution on the set* $\mathcal{X}_{t+1} = \{x : S(x) \geq \widehat{m}_{t+1}\}$.

3. ***Estimating*** c_{t+1}: *Take*

$$\widehat{c}_{t+1} = \frac{1}{N} \sum_{i=1}^{N} I_{\{S(X_i) \geq \widehat{m}_{t+1}\}} = \frac{N_{t+1}^{(e)}}{N}$$

as an estimator of c_{t+1}.

4. ***Stop:*** *If* $\widehat{m}_{t+1} = m$*, deliver the estimators* (4.9) *and* (4.10). *Else continue with the next step.*
5. ***Screening:*** *Screen the elite set; that is, remove duplicates, to obtain* $N_{t+1}^{(s)}$ *distinct points* $\{X_1^{(s)}, \ldots, X_{N_{t+1}^{(s)}}^{(s)}\}$ *uniformly distributed in* $\mathcal{X}_{t+1}$.
6. ***Splitting:*** *Reproduce (clone)* η_{t+1} *times each screened elite* $X^{(s)}$ *of the screened elite sample; that is, take* η_{t+1} *identical copies of each point.*
7. ***Sampling:*** *To each of the cloned points apply a Markov chain sampler of length* b_{t+1} *in* $\mathcal{X}_{t+1}$ *with* $g_{t+1} = \mathsf{U}(\mathcal{X}_{t+1})$ *as its stationary distribution. The length* b_{t+1} *is given by* $\eta_{t+1} b_{t+1} = \lfloor N / N_{t+1}^{(s)} \rfloor$. *Choose randomly (without replacement)* $N - N_{t+1}^{(s)} \eta_{t+1} b_{t+1}$ *of these paths and extend these with one point by applying an extra transition of the Markov chain sampler. Denote the new entire sample by* $[X]_{t+1} = \{X_1, \ldots, X_N\}$. *Note that each point in the sample is distributed uniformly on the set* $\mathcal{X}_{t+1}$.
8. ***Iterate:*** *Increase the counter* $t = t + 1$*, and repeat from step 2.*

As an illustrative example, one may consider a sum of Bernoulli random variables. Suppose we have $X_i \sim \mathsf{Ber}(0.5)$ for $i \in \{1, \ldots, n\}$ and we are interested in calculating a probability of the event $A = \{\sum_{i=1}^{n} X_i \geq m\}$ for fixed m, $m \leq n$. One should note that even for moderate n and, say, $\gamma = 0.9n$ we have a rare-event scenario. Let now $S(X) = \sum_{i=1}^{n} X_i$ be the performance function. The Gibbs sampler can be defined as follows:

Algorithm 4.3 ***Gibbs Sampler for the Sum of Bernoulli Random Variables***

- *Input: A random Bernoulli vector* $X = (x_1, \ldots, x_n)$ *where* $S(X) \geq m_t$.
- *Output: A uniformly chosen random neighbor* Y *with the same property.*

For each $i \in \{1, \ldots, n\}$ *do:*

1. *Generate* $B \sim \mathsf{Ber}(0.5)$; *set* $Y = (Y_1, \ldots, Y_{i-1}, B, x_{i+1}, \ldots, x_n)$.
2. *If* $S(Y) \geq m_t$*, set* $Y_i = B$*; otherwise* $Y_i = 1 - B$.

From this simple example, it is clear that if we repeatedly apply this algorithm on any vector X, it will eventually approach the desired event space $A = \{\sum_{i=1}^{n} X_i \geq \gamma\}$.

We conclude with a few remarks.

Remark 4.1

One of the important features of Algorithm 4.2 is that it produces points distributed nearly uniformly on each subregion $\mathcal{X}_t$. Indeed, the assumption is that in the initialization step we can generate samples that are exactly uniformly distributed on the entire space $\mathcal{X}_0$. At the subsequent iterations we generate in parallel a substantial number of independent sequences of nearly uniform points. We also make sure that they have sufficient lengths. By doing so we guarantee that our Gibbs sampler mixes rapidly and the generated points at each $\mathcal{X}_t$ are indeed distributed approximately uniform, see also a relevant discussion in Gelman and Rubin [51]. We found numerically that this holds by choosing the sample size about 10–100 times larger than dimension n and the splitting parameters ρ_t not too small, say $10^{-1} \geq \rho_t \geq 10^{-3}$. For more details on uniform sampling, see Section 4.5.

The following figures supply evidence that supports the rapid mixing properties of the Gibbs sampler. We applied the splitting algorithm to a 3-SAT problem from the SATLIB benchmark problems, consisting of $n = 75$ literals and $m = 375$ clauses. In each iteration we chose arbitrarily one of the elite points as a starting point of the Gibbs sampler of length $N = 1000$. From this sequence of points $\boldsymbol{X}_1, \ldots, \boldsymbol{X}_N$ in the subset $\mathcal{X}_t$ we constructed a time series $Y_1, \ldots, Y_N$ by summating the coordinates of each point: $Y_i = \sum_{j=1}^{n} (\boldsymbol{X}_i)_j, i = 1, \ldots, N$. Finally, we computed the autocorrelation function of this time series. Figure 4.9 shows these autocorrelation functions of the first four iterations, up to lag 20. We clearly see that in each iteration the correlation vanishes rapidly.

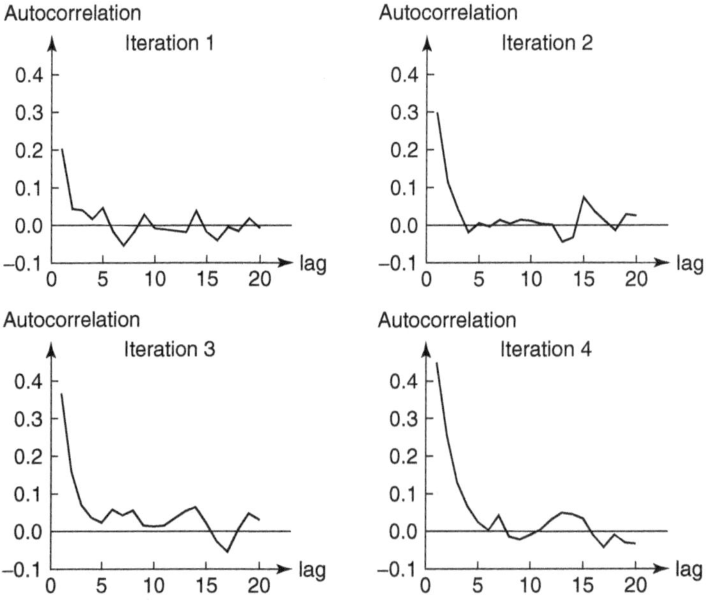

Figure 4.9 Autocorrelations of the Gibbs sampler.

Remark 4.2 *Adaptive Choice of ρ_t*

Suppose that we want to fix in advance the rarity parameters ρ_t. Typically it is not so simple, especially for combinatorial problems. The reason is that if, for example, we consider the integer constraints (1.1), then a majority of generated points will result at the same level m_t. Note that all these points are added to the elite set and as result the number of elites blows up to $N_t^{(e)} > \rho_t N$. Such a policy may lead to a situation where $N_t^{(e)} = N$. As result we obtain that $\widehat{c}_t = 1$ and the Algorithm 4.2 will be locked.

To eliminate such undesirable phenomena we propose the following adaptive procedure involving a fixed interval $(a_1, a_2) \subset (0, 1)$ for ρ_t. Initially we choose quite an arbitrary ρ_t, say $\rho_t = 0.1$, called the proposal rarity parameter; let $S_{\lceil 1-\rho_t \rceil}$ be the largest elite value of the ordered sample $S(\boldsymbol{X}_i)$, $i = 1, \ldots, N$, corresponding to the proposal ρ_t. The adaptive choice of $\rho_t \in (a_1, a_2)$, where, say $(a_1, a_2) = (0.01, 0.25)$, is executed as follows:

- Include in the elite sample all additional points satisfying $S(\boldsymbol{X}) = m_t$, provided that the number of elite samples $N_t^{(e)} \leq a_2 N$.
- Remove from the elite sample all points $S(\boldsymbol{X}) = m_t$, provided the number of elite samples $N_t^{(e)} > a_2 N$. Note that by doing so we switch from a lower elite level to a higher one. If $a_1 N \leq N_t^{(e)} \leq a_2 N$, and thus $a_1 \leq \rho_t \leq a_2$, accept $N_t^{(e)}$ as the size of the elites sample. If $N_t^{(e)} < a_1 N$, proceed sampling until $N_t^{(e)} = a_1 N$, that is until at least $a_1 N$ elite samples are obtained. The above guarantees that $\rho_t \in (a_1, a_2)$.

It follows that such adaptive ρ_t satisfying $a_1 < \rho_t \leq a_2$ prevents Algorithm 4.2 of being stopped before reaching the target level m.

Remark 4.3 *Alternative Choice of ρ*

As an alternative to Remark 4.2 we can define ρ based on the maximum value of the ordered sample $S(\boldsymbol{X}_i)$, $i = 1, \ldots, N$, denoted as $S_{\max}$, rather than on the $1 - \rho$ quantile $S_{(\lceil (1-\rho)N \rceil)}$. This is so, since the sample size N is typically larger than the number of levels m in the model. Thus, we expect that the number of $S_{\max}$ values, denoted as $\#S_{\max}$ will be greater than 1. The adaptive ρ in this case will correspond to $\rho = \#S_{\max}/N$.

As an example, consider the following two sample scenarios for the ordered values of $S_i = S(\boldsymbol{X}_i)$, $i = 1, \ldots, 9$:

i. $S_{(1)}, \ldots, S_{(9)} = 1, 1, 1, 2, 2, 2, 2, 3, 3;$

ii. $S_{(1)}, \ldots, S_{(9)} = 1, 1, 1, 1, 1, 1, 1, 2, 3.$

We have in (i) $\rho = 2/9$ and in (ii) $\rho = 1/9$. Note that because in case (ii) we have $\#S_{\max} = 1$ (that is, a single value of $S_{\max} = 3$), we can include in the elite sample all additional smaller values corresponding (in this case) to $S = 2$. By doing so we obtain $\rho = 2/9$ instead of $\rho = 1/9$.

In summary, in the alternative ρ approach we first select some target ρ value, say $\rho = 0.01$, and then accumulate all largest values of $S(\boldsymbol{X}_i)$ until we obtain

$$\frac{\text{number of accumulated largest values of } S(\boldsymbol{X}_i)}{N} \geq 0.01.$$

4.5 SAMPLING UNIFORMLY ON DISCRETE REGIONS

Here we demonstrate how, using the splitting method, one can generate a sequence of points $\boldsymbol{X}_1, \dots, \boldsymbol{X}_N$ nearly uniformly distributed on a discrete set $\mathcal{X}^*$, such as the set (1.1). By near uniformity we mean that the sample $\boldsymbol{X}_1, \dots, \boldsymbol{X}_N$ on $\mathcal{X}^*$ passes the standard Chi-square test. Thus, the goal of Algorithm 4.2 is twofold: counting the points on the set $\mathcal{X}^*$ and simultaneously generating points uniformly distributed on that set $\mathcal{X}^*$.

Note that generating a sample of points uniformly distributed on a continuous set $\mathcal{X}^*$ using a Markov chain Monte Carlo algorithm has been thoroughly studied in the literature. One of the most popular methods is the hit-and-run of Smith [53]. For recent applications of hit-and-run for convex optimization, see the work of Gryazina and Polyak [58].

We will show empirically that a simple modification of the splitting Algorithm 4.2 allows one to generate points uniformly distributed on sets like (1.1). The modification of the Algorithm 4.2 is very simple. Once it reaches the desired level $m_T = m$ we perform several more iterations, say k iterations, with the corresponding elite samples. As for previous iterations, we use here the screening, splitting, and sampling steps. Such policy will only improve the uniformity of the samples on $\mathcal{X}^*$. The number of additional steps k needed for the resulting sample to pass the Chi-square test for uniformity, depends on the quality of the original elite sample at level m, which in turn depends on the selected values ρ and b. We found numerically that the closer ρ is to 1, the more uniform is the sample. But, running the splitting Algorithm 4.2 at ρ close to 1 is clearly time consuming. Thus, there exists a trade-off between the quality (uniformity) of the sample and the number of additional iterations k required.

As an example, consider a 3-SAT problem consisting of $n = 20$ literals and $m = 80$ clauses and $|\mathcal{X}^*| = 15$ (see also Section 4.9 with the numerical results). We applied the Chi-square test for uniformity of $N = 5000$ samples for the following cases:

- $\rho = 0.05$ and no additional iterations ($k = 0$);
- $\rho = 0.5$ and no additional iterations ($k = 0$);
- $\rho = 0.5$ and one additional iteration ($k = 1$).

All three experiments passed the Chi-square test at level $\alpha = 0.05$: the observed test values were 13.709, 9.016, and 8.434, respectively, against critical value $\chi^2_{14,0.95} = 23.685$. The histogram corresponding to the last experiment ($\rho = 0.5, k = 1$) is shown in Figure 4.10.

In this way, we tested many more SAT problems for uniformity and found that typically (about 95%) the resulting samples pass the Chi-square test, provided we perform two or three additional iterations ($k = 2, 3$) on the corresponding elite sample after the algorithm has reached the desired level m.

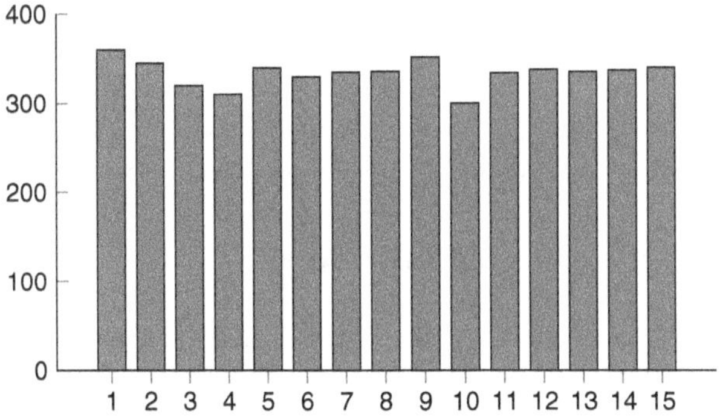

Figure 4.10 Histogram of 5,000 samples for the 3-SAT problem.

4.6 SPLITTING ALGORITHM FOR COMBINATORIAL OPTIMIZATION

In this section, we show that the splitting method is suitable for solving combinatorial optimization problems, such as the maximum cut or traveling salesman problems, and, thus, can be considered as an alternative to the standard cross-entropy and MinxEnt methods. The splitting algorithm for combinatorial optimization can be considered as a particular case of the counting Algorithm 4.2 with $|\mathcal{X}^*| = 1$. In that case, we do not need to calculate the product estimators $\widehat{c}_t$, $t = 1, \ldots, T$; thus, Step 3 of Algorithm 4.2 is omitted. The main difference reflects the stopping step.

Algorithm 4.4 ***Splitting Algorithm for Optimization***

Given the rarity parameter ρ, say $\rho \in (0.01, 0.25)$ and the sample size N, execute the following steps:

1. ***Initialize:*** *The same as in Algorithm 4.2.*
2. ***Select:*** *The same as in Algorithm 4.2 using a single ρ in all iterations.*
3. ***Stop:*** *If for some given small integer d, say $d = 4$,*

$$\widehat{m}_{t+1} = \widehat{m}_t = \cdots = \widehat{m}_{t-d} \tag{4.22}$$

   Deliver $\widehat{m}_{t+1}$ as the estimator of the optimal solution. Else go to the next step.
4. ***Screening:*** *The same as in Algorithm 4.2.*
5. ***Splitting:*** *The same as in Algorithm 4.2.*
6. ***Sampling:*** *The same as in Algorithm 4.2.*
7. ***Iterate:*** *The same as in Algorithm 4.2.*

4.7 ENHANCED SPLITTING METHOD FOR COUNTING

Here we consider two enhanced estimators for the adaptive splitting method for counting. They are called the *direct* and *capture-recapture* estimators, respectively.

4.7.1 Counting with the Direct Estimator

This estimator is based on *direct* counting of the number of screened samples obtained immediately after crossing the level m. This is the reason that it is called the direct estimator. It is denoted $\widehat{|\mathcal{X}^*|}_{\text{dir}}$, and is associated with the uniform distribution $g^*(\boldsymbol{x}, m)$ on $|\mathcal{X}^*|$. We found numerically that $\widehat{|\mathcal{X}^*|}_{\text{dir}}$ is extremely useful and very accurate as compared to $\widehat{|\mathcal{X}^*|}$ of the Algorithm 4.2, but is applicable only for counting where $|\mathcal{X}^*|$ is not too large. Specifically, $|\mathcal{X}^*|$ should be less than the sample size N. Note, however, that counting problems with small values $|\mathcal{X}^*|$ are the most difficult ones and in many counting problems one is interested in the cases where $|\mathcal{X}^*|$ does not exceed some fixed quantity, say $\mathcal{N}$. Clearly, this is possible only if $N \geq \mathcal{N}$. It is important to note that $\widehat{|\mathcal{X}^*|}_{\text{dir}}$ is typically much more accurate than its counterpart, the standard estimator $\widehat{|\mathcal{X}^*|} = \widehat{\ell}|\mathcal{X}|$ based on the splitting Algorithm 4.2. The reason is that $\widehat{|\mathcal{X}^*|}_{\text{dir}}$ is obtained directly by counting all distinct values of $\boldsymbol{X}_i$, $i = 1, \dots, N$ satisfying $S(\boldsymbol{X}_i) \geq m$, that is, it can be written as

$$\widehat{|\mathcal{X}^*|}_{\text{dir}} = \sum_{i=1}^{N_T^{(s)}} I_{\{S(X_i^{(s)}) \geq m\}} = |[\boldsymbol{X}^{(s)}]_T|, \tag{4.23}$$

where $[\boldsymbol{X}^{(s)}]_T$ is the sample obtained by screening the elites when Algorithm 4.2 actually satisfied the stopping criterion at Step 4. Again, the estimating Step 3 is omitted; and conveniently Step 4 (stopping) and Step 5 (screening) are swapped.

Algorithm 4.5 ***Direct Algorithm for Counting***

- *Input: The counting problem* (4.2)*; the splitting parameters* η_t, $t = 1, \dots, T-1$*; the rarity parameters* ρ_t, $t = 1, \dots, T-1$*; the parameters* a_1 *and* a_2*, say* $a_1 = 0.01$ *and* $a_2 = 0.25$*, such that* $\rho_t \in (a_1, a_2)$ *as in Remark 4.2; the sample size* N.
- *Output: Estimator* (4.23).

1. ***Initialization:*** *Same as in Algorithm 4.2.*
2. ***Select:*** *Same as in Algorithm 4.2.*
3. ***Screening:*** *Same as in Algorithm 4.2.*
4. ***Stop:*** *If* $\widehat{m}_{t+1} = m$*, deliver* (4.23) *as an estimator of* $|\mathcal{X}^*|$*. Else continue with next step.*
5. ***Splitting:*** *Same as in Algorithm 4.2.*

6. ***Sampling:*** *Same as in Algorithm 4.2.*
7. ***Iterate:*** *Same as in Algorithm 4.2.*

To increase further the accuracy of $\widehat{|\mathcal{X}^*|}_{\text{dir}}$ we might execute one more iteration with a larger sample (more clones in Step 5 and/or longer Markov paths in Step 6).

Note that there is no need to calculate $\widehat{c}_t$ in Algorithm 4.5 at any iteration. Note also that the counting Algorithm 4.5 can be readily modified for combinatorial optimization, since an optimization problem can be viewed as a particular case of counting, where the counted quantity $|\mathcal{X}^*| = 1$.

4.7.2 Counting with the Capture–Recapture Method

In this section, we discuss how to combine the classic capture-recapture (CAP-RECAP) method with the splitting Algorithm 4.2. First we present the classical capture-recapture algorithm in the literature.

4.7.2.1 Capture-Recapture Method

Originally, the capture-recapture method was used to estimate the size, say M, of some unknown population on the basis of two independent samples, each taken without replacement. To see how the CAP-RECAP method works, consider an urn model with a total of M identical balls. Denote by N_1 and N_2 the sample sizes taken at the first and second draws, respectively. Assume, in addition, that

- The second draw takes place after all N_1 balls have been returned to the urn.
- Before returning the N_1 balls, each is marked, say, painted with a different color.

Denote by R the number of balls from the first draw that reappear in the second. Then a biased estimate $\widetilde{M}$ of M is

$$\widetilde{M} = \frac{N_1 N_2}{R}. \tag{4.24}$$

This is based on the observation that $N_2/M \approx R/N_1$. Note that the name capture-recapture was borrowed from a problem of estimating the animal population size in a particular area on the basis of two visits. In this case, R denotes the number of animals captured on the first visit and recaptured on the second.

A slightly less biased estimator of M is

$$\widehat{M} = \frac{(N_1 + 1)(N_2 + 1)}{(R + 1)} - 1. \tag{4.25}$$

See Seber [113] for an analysis of its bias. Furthermore, defining the statistic

$$V = \frac{(N_1 + 1)(N_2 + 1)(N_1 - R)(N_2 - R)}{(R + 1)^2 (R + 2)},$$

Seber [113] shows that

$$\mathbb{E}[V] \sim \mathbb{V}\text{ar}[\widehat{M}]\ (1 + \mu^2 e^{-\mu}),$$

where

$$\mu = \mathbb{E}[R] = N_1 N_2 / M,$$

so that V is an approximately unbiased estimator of the variance of $\widehat{M}$.

4.7.2.2 Splitting Algorithm Combined with Capture–Recapture

Application of the CAP-RECAP to counting problems is straightforward. The target is to estimate the unknown size $M = |\mathcal{X}^*|$. Consider the last iteration T of the adaptive splitting Algorithm 4.2; that is, when we have generated in Step 2 the set $[\boldsymbol{X}^{(e)}]_T$ of elite points at level m (points in the target set $\mathcal{X}^*$).

Then, we perform the following steps:

1. Skip the stopping Step 4 (of Algorithm 4.2).
2. Screen the elites (Step 5).
3. Execute splitting Step 6 and sampling Step 7 (of Algorithm 4.2) with a sample size $N_1^{(\text{cap})}$.
4. Remove the duplicates.
5. Record the resulting set of N_1 distinct points.
6. Execute Steps 2–5 above a second time, starting from the same elite set $[\boldsymbol{X}]_T^{(e)}$; the sample size in Step 5 is taken to be $N_2^{(\text{recap})}$. Let N_2 be the number of distinct points.
7. Count the number R of distinct points that occur in both runs and deliver either estimator the (4.24) or (4.25).

The resulting algorithm can be written as

Algorithm 4.6 ***Capture-Recapture Algorithm for Counting***

- *Input: the counting problem* (4.2)*; sample sizes* $N, N_1^{(\text{cap})}, N_2^{(\text{recap})}$*; rarity parameters* ρ_t*; set automatically the splitting parameters* $\eta_t \equiv 1$.
- *Output: counting estimator* (4.24) *or* (4.25).

1. ***Initialize:*** *Set a counter* $t = 0$. *Generate a sample* $[\boldsymbol{X}]_0 = \{\boldsymbol{X}_1, \ldots, \boldsymbol{X}_N\}$ *uniformly on* $\mathcal{X}_0$.
2. ***Select:*** *Compute level* $\widehat{m}_{t+1}$ *as the* $(1 - \rho_{t+1})$ *quantile of the ordered statistics values of* $S(\boldsymbol{X}_1), \ldots, S(\boldsymbol{X}_N)$. *Determine the elite sample, that is, the largest subset of* $[\boldsymbol{X}]_t$ *consisting of points for which* $S(\boldsymbol{X}_i) \geq \widehat{m}_{t+1}$. *Suppose that* $N_{t+1}^{(e)}$ *is the size of this subset, and denote its points by* $\boldsymbol{X}_j^{(e)}$*; thus,*

$$[\boldsymbol{X}^{(e)}]_{t+1} = \{\boldsymbol{X}_1^{(e)}, \ldots, \boldsymbol{X}_{N_{t+1}^{(e)}}^{(e)}\}.$$

3. ***Screening:*** *Screen the elite set; that is, remove duplicates to obtain* $N_{t+1}^{(s)}$ *distinct points* $\{\boldsymbol{X}_1^{(s)}, \ldots, \boldsymbol{X}_{N_{t+1}^{(s)}}^{(s)}\}$, *uniformly distributed in* $\mathcal{X}_{t+1}$.
4. ***Stop:*** *If* $\widehat{m}_{t+1} = m$, *set* $T = t + 1$ *and go to Step 7; otherwise continue with the next step.*
5. ***Sampling:*** *To each of the screened elite points, apply a Markov chain sampler of length* b_{t+1} *in* $\mathcal{X}_{t+1}$ *with* $g_{t+1} = \mathsf{U}(\mathcal{X}_{t+1})$ *as its stationary distribution. The length* b_{t+1} *is given by* $b_{t+1} = \lceil N/N_{t+1}^{(s)} \rceil$. *Choose randomly (without replacement)* $N - N_{t+1}^{(s)} b_{t+1}$ *of these paths and extend these with one point by applying an extra transition of the Markov chain sampler. Denote the new entire sample by* $[\boldsymbol{X}]_{t+1} = \{\boldsymbol{X}_1, \ldots, \boldsymbol{X}_N\}$.
6. ***Iterate:*** *Increase the counter* $t = t + 1$, *and repeat from step 2.*
7. ***Capture:*** *Let* $b_T(1) = \lfloor N_1^{(\text{cap})} / N_T^{(s)} \rfloor$. *For all* $i = 1, 2, \ldots, N_T^{(s)}$, *starting at the i-th screened elite point, run a Markov chain of length* $b_T(1)$ *in* $\mathcal{X}_T$ *with* $g_T = \mathsf{U}(\mathcal{X}_T)$ *as its stationary distribution. Extend* $N_1^{(\text{cap})} - N_T^{(s)} b_T(1)$ *randomly chosen sample paths with one point. Screen out duplicates to obtain a set* $[\boldsymbol{X}]_1^{(\text{cap})}$ *of* N_1 *distinct points in* $\mathcal{X}^*$.
8. ***Recapture:*** *Repeat step 7 with* $b_T(2) = \lfloor N_2^{(\text{recap})} / N_T^{(s)} \rfloor$. *After screening out, the remaining set is* $[\boldsymbol{X}]_2^{(\text{recap})}$ *of* N_2 *distinct points in* $\mathcal{X}^*$.
9. ***Capture-Recapture:*** *Compute the number R of points in* $[\boldsymbol{X}]_1^{(\text{cap})} \cap [\boldsymbol{X}]_2^{(\text{recap})}$.
10. ***Estimating:*** *Deliver estimator* (4.24) *or* (4.25).

In Section 4.9, we compare the numerical results of the adaptive splitting Algorithm 4.2 with the capture–recapture Algorithm 4.6. As a general observation we found that the performances of the two counting estimators depend on the size of the target set $\mathcal{X}^*$. In particular, when we keep the sample N limited to 10,000, then for $|\mathcal{X}^*|$ sizes up to 10^6 the CAP-RECAP estimator (4.25), denoted as $\widehat{|\mathcal{X}^*|}_{\text{cap}}$ is more accurate than the adaptive splitting estimator $\widehat{|\mathcal{X}^*|}$ in (4.10), that is,

$$\mathbb{V}\text{ar}[\widehat{|\mathcal{X}^*|}_{\text{cap}}] \leq \mathbb{V}\text{ar}[\widehat{|\mathcal{X}^*|}].$$

However, for larger target sets, say with $|\mathcal{X}^*| > 10^6$, we propose using the splitting algorithm because the capture–recapture method performs poorly.

4.8 APPLICATION OF SPLITTING TO RELIABILITY MODELS

4.8.1 Introduction

Consider a connected undirected graph $G = G(V, E)$ with n vertices and m edges, where each edge has some probability p of failing. What is the probability that G becomes disconnected under random, independent edge failures? This problem can be generalized by allowing a different failure probability for each edge. It is well known that this problem is #P-hard, even in the special case $p = 1/2$.

Although approximation [18] and bounding [5, 6] methods are available, their accuracy and scope are very much dependent on the properties (such as size and topology) of the network. For large networks, estimating the reliability via simulation techniques becomes desirable. Due to the computational complexity of the exact determination, a Monte Carlo approach is justified. However, in highly reliable systems such as modern communication networks, the probability of network failure is a rare-event probability and naive methods such as the Crude Monte Carlo are impractical. Various variance reduction techniques have been developed to produce better estimates. Examples include a control variate method [47], importance sampling [22, 59], and graph decomposition approaches such as the recursive variance reduction algorithm and its refinements [21, 23]. For a survey of these methods, see [20].

For network unreliability, Karger [64] presents an algorithm and shows that its complexity belongs to the so-called fully polynomial randomized approximation schemes (see Appendix section A.3.2 for a definition). His approach can be described as follows.

Assume for simplicity that each edge fails with equal probability q. Let c be the size of a minimum cut in G; that is, the smallest cut in the graph having exactly c edges. An important observation of Karger is that if all edges of the cut fail, then the graph becomes disconnected and the network unreliability (failure probability), denoted by q_{fail}, satisfies $q_{\text{fail}} \geq q^c$. If q^c is bounded by some reasonable polynomial, say $q^c \geq \frac{1}{n^4}$, then the naive Monte Carlo will do the job. Otherwise, q^c will be very small and, thus, we are forced to face a rare event situation. Karger discusses the following issues:

- The failure probability of a cut decreases exponentially with the number of edges in the cut (so he suggests considering only relatively small cuts).
- There exists a polynomial number of such relatively small cuts, and they can be enumerated in polynomial time.
- To estimate q_{fail} to within $1 \pm \varepsilon$ with high probability, it is sufficient to perform $\mathcal{O}((\log\ n)/(\varepsilon^2 q_{\text{fail}}))$ trials.
- Combining the well-known DNF counting algorithm of Karp and Luby [65], one can construct easily a polynomial-sized DNF formula whose probability of being satisfied is precisely the desired one. The algorithm of Karger estimates the failure probability of the graph to within $1 \pm \varepsilon$ in $\mathcal{O}(cn^4/\varepsilon^3)$ time.

The *permutation Monte Carlo* (PMC) and the *turnip* methods of Elperin, Gertsbakh, and Lomonosov [41, 42] are the most successful breakthrough methods for networks reliability. Both methods are based on the so-called *basic Monte Carlo* (BMC) algorithm for computing the sample performance $S(\boldsymbol{Y})$ (see [52] and Algorithm 4.7). It is shown in [41, 42] that the complexity of both PMC and turnip is $\mathcal{O}(m^2)$, where m is the number of edges in the network.

Botev et al. [14] introduce a fast splitting algorithm based on efficient implementation of data structures using minimal spanning trees. This allows one to speed-up substantially the Gibbs sampling, while estimating the sample reliability function $S(\boldsymbol{Y})$. The Gibbs sampler is implemented online into the BMC Algorithm 4.7 (see Section 4.8.3). Although numerically the splitting method in [14] performs exceptionally well compared with its standard counterpart (see Algorithms 16.7–16.9 in Kroese et al. [73]), there is no formal proof regarding the speed-up of the former relative to the latter. Extensive numerical results support the conjecture that its complexity is the same as in [41, 42], that is, $\mathcal{O}(m^2)$.

The recent work of Rubinstein et al. [109] deals with further investigation of networks' reliability. In particular, they consider the following three well-known methods: cross-entropy, splitting, and PMC. They show that combining them with the so-called *hanging edges Monte Carlo* (HMC) algorithm of Lomonosov [43] instead of the conventional BMC algorithm obtains a speed-up of order $m/\ln\ (m)$. The speed-up is due to the fast computation of the sample reliability function $S(\boldsymbol{Y})$. As a consequence, the earlier suggested cross-entropy algorithm in Kroese et al. [59], the conventional splitting one in Kroese et al. [73], and the PMC algorithm of Elperin et al. [41] for network reliability (all based on the BMC Algorithm 4.7) can be sped up by a factor of $m/\ln\ (m)$. In particular, the following is shown:

- The theoretical complexity of the new HMC-based PMC algorithm becomes $\mathcal{O}(m \ln\ m)$ instead of $\mathcal{O}(m^2)$.
- The empirical complexity of the new HMC-based cross-entropy algorithm becomes $\mathcal{O}(m \ln\ m)$ instead of $\mathcal{O}(m^2)$.
- The empirical complexity of the new HMC-based splitting algorithm becomes $\mathcal{O}(m^2)$ instead of $\mathcal{O}(m^2 \ln\ m)$.

An essential feature of all three algorithms in [109] is that they retain their original form, because the improved counterparts merely implement the HMC as a subroutine.

In spite of the nice theoretical and empirical complexities of the new HMC-based PMC and cross-entropy algorithms, they have limited application. The reason is that both cross-entropy and PMC algorithms are numerically unstable for large m. Consequently, the cross-entropy and PMC algorithms are able to generate stable reliability estimators for m up to several hundred nodes, while the splitting and the turnip algorithms (in [14, 109], and [41, 42], respectively, all with complexities $\mathcal{O}(m^2)$) are able to generate accurate reliability estimators for tens of thousands of nodes.

Here we present a splitting algorithm based on the BMC Algorithm 4.7, as an alternative to the methods of [73] and [109]. Similar to [14] its empirical complexity is $\mathcal{O}(m^2)$. We believe that the advantage of the current version as compared to the one in [14] is the simplicity of implementation of the Gibbs sampler for generating $S(\boldsymbol{Y})$.

4.8.2 Static Graph Reliability Problem

Consider an undirected graph $G = G(V, E)$ with a set of nodes V and a set of edges E. Each edge can be either operational or failed. The configuration of the system can be denoted by $\boldsymbol{x} = (x_1, \ldots, x_m)$, where m is the number of edges, $x_i = 1$ if edge i is operational, and $x_i = 0$ if it is failed. A subset of nodes $V_0 \subset V$ is selected a priori and the system (or graph) is said to be operational in the configuration $\boldsymbol{x}$ if all nodes in V_0 are connected to each other by at least one path of operational edges only. We denote this by defining the structure function,

$$\varphi : \{0, 1\}^m \to \{0, 1\},$$

by $\varphi(\boldsymbol{x}) = 1$ when the system is operational in configuration $\boldsymbol{x}$, and $\varphi(\boldsymbol{x}) = 0$ otherwise.

Suppose that edge i is operational with probability r_i, so that the system configuration is a random vector $\boldsymbol{X} = (X_1, \ldots, X_m)$, where $\mathbb{P}(X_i = 1) = r_i = 1 - \mathbb{P}(X_i = 0)$ and the X_i's are independent. The goal is to estimate $r = \mathbb{P}(\varphi(\boldsymbol{X}) = 1)$, the reliability of the system.

The crude Monte Carlo reliability estimator is given as

$$\widehat{r} = \frac{1}{N} \sum_{k=1}^{N} \varphi(\boldsymbol{X}_k), \tag{4.26}$$

where $\boldsymbol{X}_1, \ldots, \boldsymbol{X}_N$ are iid realizations from a multivariate Bernoulli distribution such that $\mathbb{P}(X_i = 1) = r_i$ for all i. For highly reliable systems, the unreliability $\bar{r} = 1 - r$ is very small, so that we have a rare-event situation, and N has to be prohibitively large to get a meaningful estimator. Such rare-event situations call for efficient sampling strategies. In this section, we shall use the ones based on cross-entropy, splitting, and the PMC method, all of which rely on the graph evolution approach described next.

4.8.2.1 Dynamic Graph Reliability Problem Formulation

In our context, a rare event occurs when the network is nonoperational. We consider an artificial state that contains more information than the binary vector based on [41]. The idea is to transform the static model into a dynamic one by assuming that all edges start in the failed state at time 0 and that the i-th edge is repaired after a random time Y_i, whose distribution is chosen so that the probability of repair at time 1 is the reliability of the edge. In particular, we assume that the Y_i's have a continuous cumulative distribution function $F_i(y) = \mathbb{P}(Y_i \leq y)$ for $y \geq 0$, with $F_i(0) = 0$ and $F_i(1) = r_i$. The reliability of the graph is the probability that it is operational at time 1, and the crude Monte Carlo algorithm can be reformulated as follows:

> Generate the vector of repair times and check if the graph is operational at time 1; repeat this N times independently, and estimate the reliability by the proportion of those graphs that are operational.

This formulation has the advantage that, given the vector $\boldsymbol{Y} = (Y_1, \ldots, Y_m)$, we can compute at what time the network becomes operational and use this real number as a measure of our closeness to the rare event.

Assuming that the Y_i's are independent, they are easy to generate by inverting the F_i's. We define random variables $X_i(s) = I_{\{Y_i \le s\}}$ for all $s \ge 0$, where I is the indicator function, and the associated random configuration $\boldsymbol{X}(s) = (X_1(s), \ldots, X_m(s))$. These variables give the status of edge i at time s, namely, if $X_i(s) = 1$, the edge is operational at time s. Note that for the original Bernoulli variable indicating whether edge i is operational, we have $X_i \stackrel{\mathcal{D}}{=} X_i(1)$. In this way we get also a random dynamic structure function $\varphi(\boldsymbol{X}(s))$, $s \ge 0$, for which $\varphi(\boldsymbol{X}) \stackrel{\mathcal{D}}{=} \varphi(\boldsymbol{X}(1))$. The interpretation is that the edges become operational one by one at random times and we are interested in the reliability at time 1. This is referred to as the *construction process* in [110]. Define the performance function

$$S(\boldsymbol{Y}) = \inf\{s \ge 0 : \varphi(\boldsymbol{X}(s)) = 1\}$$

as the first time when the graph becomes operational. Hence, the value equals one of the repair times Y_i, namely the one that swaps the graph from non-operational to operational. Another point of view is to consider $S(\boldsymbol{Y})$ as a measure of closeness of $\boldsymbol{Y}$ to reliability: $S(\boldsymbol{Y}) \le 1$ means that the system is operational at time 1, thus $(S(\boldsymbol{Y}) - 1)^+$ measures the distance from it. Both the cross-entropy [59] and the splitting method [73] exploit this feature.

These methods are designed to estimate the nonoperational probability (or unreliability) $\mathbb{P}(S(\boldsymbol{Y}) > 1)$. Ideally, we would like to sample $\boldsymbol{Y}$ (repeatedly) from its distribution conditional on $S(\boldsymbol{Y}) > 1$. To achieve this, we first partition the interval [0, 1] in fixed levels

$$0 = m_0 < m_1 < \cdots < m_\tau = 1,$$

and use them as intermediate stepping stones on the way to sampling $\boldsymbol{Y}$ conditional on $S(\boldsymbol{Y}) > 1$ as follows.

At stage t of the algorithm, for $t \ge 1$, we have a collection of states $\boldsymbol{Y}$ from their distribution conditional on $\{S(\boldsymbol{Y}) > m_{t-1}\}$. For each of them, we construct a Markov chain that starts with that state and evolves by resampling some of the edge repair times under their distribution conditional on the value S remaining larger than m_{t-1}. While running all those chains for a given number of steps, we collect all the visited states $\boldsymbol{Y}$ whose value is above m_t, and use them for the next stage. By discarding the repair times below m_t and starting new chains from those above m_t, and repeating this at each stage, we favor the longer repair times and eventually obtain a sufficiently large number of states with values above 1. Based on the number of steps at each stage, and the proportion of repair times that reach the final level, we obtain an unbiased estimator of $\mathbb{P}(S(\boldsymbol{Y}) > 1)$, which is the graph unreliability, see [41].

This reformulation with the repair-time variables Y_i enables us to consider the reliability at any time $s > 0$: the configuration is $\boldsymbol{X}(s)$ and the graph is operational if and only if $\varphi(\boldsymbol{X}(s)) = 1$. Once the repair-time variables Y_i are generated, we can compute from them an estimator of the reliability for any $s > 0$.

In Elperin et al. [41, 42] and Gertsbakh and Shpungin [52], each repair distribution F_i is taken as exponential with rate $\lambda_i = -\ln(1 - r_i)$ (denoted $\mathsf{Exp}(\lambda_i)$). Then

$$\mathbb{P}(Y_i \leq s) = 1 - e^{-\lambda_i s} = 1 - (1 - r_i)^s$$

and $\mathbb{P}(Y_i \leq 1) = r_i$. We denote the corresponding joint pdf by $f(\boldsymbol{s}, \boldsymbol{\lambda})$.

4.8.3 BMC Algorithm for Computing $S(Y)$

Recall that, in terms of the vector $\boldsymbol{Y}$, the graph unreliability can be written as $\bar{r} = \mathbb{P}(S(\boldsymbol{Y}) > 1)$ and using the crude Monte Carlo, we would estimate $\bar{r}$ via

$$\widehat{\bar{r}}_{\text{CMC}} = \frac{1}{N} \sum_{i=1}^{N} I_{\{S(\boldsymbol{Y}_i) > 1\}}, \quad \boldsymbol{Y}_1, \ldots, \boldsymbol{Y}_N \overset{\text{IID}}{\sim} f(\boldsymbol{y}, \boldsymbol{\lambda}).$$

It is well known that direct evaluation of $S(\boldsymbol{Y})$ by using the minimal cuts and minimal path sets of the network is impractical. Instead, it is common to resort to the following simple algorithm [52, 59, 73], which uses a breadth-first search procedure [32].

Algorithm 4.7 ***Basic Monte Carlo Algorithm for Computing*** $S(\boldsymbol{Y})$

Given $\boldsymbol{Y}$ *and a set of terminal nodes of size* K, *set* $b = 1$ *and execute the following steps.*

1. *Let* $\boldsymbol{\pi} = (\pi_1, \ldots, \pi_m)$ *be the permutation of the edges* $1, \ldots, m$ *obtained from* $\boldsymbol{Y}$ *such that*
$$Y_{\pi_1} < Y_{\pi_2} < \cdots < Y_{\pi_m}.$$
2. *Consider the network* G *in which the edges* $\pi_1, \ldots, \pi_b$ *are operational and the rest,* $\pi_{b+1}, \ldots, \pi_m$, *are failed.*
3. *Use the breadth-first search algorithm to check if the network is operational. If so, denote by* τ *the final number* b *corresponding to the operational network, stop and go to Step 4; otherwise, increment* $b = b + 1$ *and repeat from Step 2.*
4. *Output* $S(\boldsymbol{Y}) = Y_{\pi_\tau}$ *as the time at which the network becomes operational.*

The stopping random value of τ corresponding to the operational network is called the *critical number*.

Remark 4.4

In the original BMC version [52], the terminal connectivity (see Step 3 of Algorithm 4.7) was established using the Kruskal algorithm with complexity $\mathcal{O}(m \ln(n))$. The corresponding complexity of the entire algorithm was $\mathcal{O}(m^2 \ln(n))$ [52]. One can speed up

Algorithm 4.7 by using the bisection search for finding the critical number τ instead of searching it sequentially starting from $b = 1$. By doing so, the complexity of this enhanced breadth-first search version reduces to $\mathcal{O}(m)$ and that of Algorithm 4.7 reduces to $\mathcal{O}(m^2)$.

4.8.4 Gibbs Sampler

To estimate the unreliability $\bar{r}$ we use the standard splitting algorithm on a sequence of adaptive levels

$$m_1 \leq m_2 \leq \cdots \leq m_{T-1} \leq m_T = 1.$$

As usual, for each level m_t we employ a Gibbs sampler. Clearly, the complexity of the splitting algorithm depends on the complexity of the Gibbs sampler. We present here two version of the Gibbs sampler, the *standard* one, which was used earlier for different estimation and counting problems [105], and the new faster one, called an *enhanced* Gibbs sampler, which is specially designed for reliability models.

Algorithm 4.8 ***Standard Gibbs Sampler for Network Reliability***

- *Input: A time* $s > 0$*; a graph* $G = G(V, E)$*, with* $|E| = m$*; repair times* $\tilde{\boldsymbol{y}} = (\tilde{y}_1, \ldots, \tilde{y}_m) \in [0, \infty)^m$ *such that* $S(\tilde{\boldsymbol{y}}) > s$*.*
- *Output: A random neighbor* $\boldsymbol{Y} \in [0, \infty)^m$ *such that* $S(\boldsymbol{Y}) > s$*.*

For each $i \in \{1, \ldots, m\}$ *do:*

1. *Generate* $Y_i \sim \mathsf{Exp}(\lambda_i)$*; set* $\boldsymbol{Y} = (Y_1, \ldots, Y_i, \tilde{y}_{i+1}, \ldots, \tilde{y}_m)$*.*
2. *If* $S(\boldsymbol{Y}) \leq s$*, set* $Y_i = s + Y_i$*.*

This algorithm exploits two properties, which ensures its correctness. The first property is that, given any repair times $\boldsymbol{y} = (y_1, \ldots, y_m)$, the value $S(\boldsymbol{y}) = y_j$, where y_j is the smallest of all repair times that make the system operational. Suppose that $S(\boldsymbol{y}) > s$, and choose an arbitrary edge e_i. Suppose that we may alter its repair time randomly to set it Y_i. Then there are two possible scenarios.

a. Either edge e_i is noncritical in the sense that, whatever the new repair time Y_i will be, the perfomance value remains larger than s.

b. Or edge e_i is critical in the sense that setting Y_i too small, the value becomes less than s. In the latter case, surely when $Y_i > s$, the value also remains larger. Then we call the second property, which is the memoryless property of the exponential distribution.

The Gibbs sampler Algorithm 4.8 is quite time consuming because the evaluation of $S(\boldsymbol{y})$ is so. Below we propose a faster option, called the *enhanced Gibbs sampler*, which avoids explicit calculation of $S(\boldsymbol{y})$ during Gibbs sampler execution. It is based on similar observations made above. Consider again the situation that

$S(\mathbf{y}) = y_j > s$ for some repair time y_j and that we change the repair time of edge e_i randomly.

- If already $y_i \le s$, then we conclude that edge e_i is noncritical, so we can generate any repair time $Y_i \sim \mathsf{Exp}(\lambda_i)$.
- If $y_i > s$, then we run a breadth-first search procedure [32] in order to verify whether edge e_i is critical or noncritical. The breadth-first search procedure will operate on a subgraph $G' = G(V', E')$ induced by edges operational before time s. If e_i is indeed critical, we will generate $Y_i \sim s + \mathsf{Exp}(\lambda_i)$.

The enhanced Gibbs sampler can be written as follows.

Algorithm 4.9 ***Enhanced Gibbs Sampler for Network Reliability***

- *Input: A time $s > 0$; a graph $G = G(V, E)$, with $|E| = m$; repair times $\tilde{\mathbf{y}} = (\tilde{y}_1, \ldots, \tilde{y}_m) \in [0, \infty)^m$ such that $S(\tilde{\mathbf{y}}) > s$.*
- *Output: A random neighbor $\mathbf{Y} \in [0, \infty)^m$ such that $S(\mathbf{Y}) > s$.*

For each $i \in \{1, \ldots, m\}$ do:

1. *Generate $Y_i \sim \mathsf{Exp}(\lambda_i)$; set $\mathbf{Y} = (Y_1, \ldots, Y_i, \tilde{y}_{i+1}, \ldots, \tilde{y}_m)$.*
2. *If $\tilde{y}_i > s$:*
   - **a.** *Determine $E' = \{e_j \in E : (\mathbf{Y})_j \le s\}$ and $V' = \{v \in V : (u, v) \in E'\}$.*
   - **b.** *Check terminal connectivity by applying the breadth-first search procedure to the subgraph $G' = G(V', E')$.*
   - **c.** *If the terminals are connected set $Y_i = s + Y_i$.*

Taking into account that for each random variable Y_i the most expensive operation is breadth-first search procedure on G', which takes at most m units of time because $G' = G(V', E')$ is clearly a subgraph of the original $G = G(V, E)$ and so $|E'| \le |E| = m$, we conclude that the overall complexity of the enhanced Gibbs sampler Algorithm 4.9 is $\mathcal{O}(m^2)$.

4.9 NUMERICAL RESULTS WITH THE SPLITTING ALGORITHMS

We present here numerical results with the proposed algorithms for counting and optimization. A large collection of instances (including real-world) is available on the following sites:

- http://people.brunel.ac.uk/~mastjjb/jeb/orlib/scpinfo.html
- http://www.nlsde.buaa.edu.cn/~kexu/benchmarks/set-benchmarks.htm
- http://www.mat.univie.ac.at/~neum/glopt/test.html

- http://www.caam.rice.edu/~bixby/miplib/miplib.html
- Multiple-knapsack are given on
  - http://hces.bus.olemiss.edu/tools.html
  - http://elib.zib.de/pub/Packages/mp-testdata/ip/sac94-suite/index.html
  - http://www.diku.dk/~pisinger/codes.html
- SAT problems are given on the SATLIB website, www.satlib.org, and
  - http://www.satcompetition.org
  - http://fmv.jku.at/sat-race-2006/downloads.html

4.9.1 Counting

4.9.1.1 SAT Problem

We present data from experiments with two different 3-SAT models:

- Small size; 20 × 80 (meaning $n = 20$ variables and $m = 80$ clauses); exact count $|\mathcal{X}^*| = 15$.
- Moderate size; 75 × 325; exact count $|\mathcal{X}^*| = 2258$.

We shall use the following notation.

Notation 4.1

For iteration $t = 1, 2, \ldots$

- $N_t^{(e)}$ and $N_t^{(s)}$ denote the actual number of elites and the number after screening, respectively;
- m_t^* and m_{*t} denote the upper and the lower elite levels reached, respectively (the m_{*t} levels are the same as the m_t levels in the description of the algorithm);
- ρ_t, η_t, b_t are the adaptive rarity, splitting, and burn-in parameters, respectively;
- $\widehat{c}_t = N_t^{(e)}/N$ is the estimator of the t-th conditional probability;
- (intermediate) product estimator after the t-th iteration $\widehat{|\mathcal{X}_t^*|} = |\mathcal{X}| \prod_{i=1}^t \widehat{c}_i$;
- (intermediate) direct estimator after the t-th iteration $\widehat{|\mathcal{X}_t^*|}_{\text{dir}}$, which is obtained by counting directly the number of distinct points satisfying all clauses.
- RE denotes the estimated relative errors defined in (2.3) in Chapter 2.
- CPU report computing time of the algorithm. All computations were executed on the same machine (PC Pentium E5200 4GB RAM).

Table 4.1 Dynamics of Algorithm 4.2 for 3-SAT (20 × 80) model

t	$\widehat{\vert\mathcal{X}_t^*\vert}$	$\widehat{\vert\mathcal{X}_t^*\vert}_{\text{dir}}$	$N_t^{(e)}$	$N_t^{(s)}$	m_t^*	m_{*t}	$\widehat{c}_t$
1	1.59E+05	0	152	152	78	56	0.152
2	3.16E+04	0	198	198	78	74	0.198
3	8.84E+03	0	280	276	79	76	0.280
4	1.78E+03	3	201	190	80	77	0.201
5	229.11	6	129	93	80	78	0.129
6	15.580	15	68	15	80	79	0.068
7	15.580	15	1000	15	80	80	1.000

4.9.1.2 First Model: 3-SAT with Instance Matrix A = (20 × 80)

This instance of the 3-SAT problem consists of $n = 20$ variables and $m = 80$ clauses. The exact count is $|\mathcal{X}^*| = 15$. A typical dynamics of a run of the adaptive splitting Algorithm 4.2 with $N = 1{,}000$, $\rho_t = 0.1$, and $b_t = 1$ (single burn-in) is given in Table 4.1.

We ran the algorithm 10 times with $N = 1{,}000, \rho_t = 0.1$, and $b_t = 1$. The average performance was

$$\widehat{|\mathcal{X}^*|} = 15.504, \ \text{RE} = 0.118, \ \text{CPU} = 5.153.$$

The direct estimator $\widehat{|\mathcal{X}^*|}_{\text{dir}}$ gave always the exact result $|\mathcal{X}|^* = 15$.

4.9.1.3 Second Model: 3-SAT with Instance Matrix A = (75 × 325)

Our next model is a 3-SAT with an instance matrix $A = (75 \times 325)$ taken from www.satlib.org. The exact value is $|\mathcal{X}^*| = 2258$. Table 4.2 presents the dynamics of the adaptive Algorithm 4.2 for this problem.

We ran the splitting algorithm 10 times, using sample size $N = 10{,}000$ and parameters $\rho_t \equiv 0.1$ and $b_t = \eta_t$ for all iterations. We obtained

$$\widehat{|\mathcal{X}^*|} = 2231.8, \ \text{RE} = 0.072, \ \text{CPU} = 688.6.$$

We also implemented the direct Algorithm 4.5 after the last iteration of the adaptive splitting using $N = 100{,}000$. We obtained

$$\widehat{|\mathcal{X}^*|}_{\text{dir}} = 2210.4, \ \text{RE} = 0.009.$$

Table 4.3 presents a comparison of the performances of the adaptive splitting estimator $\widehat{|\mathcal{X}^*|}$ and its counterpart, the CAP-RECAP estimator, denoted by $\widehat{|\mathcal{X}^*|}_{\text{cap}}$ obtained from the capture-recapture Algorithm 4.6. The comparison was done for the same random 3-SAT model with the instance matrix $A = (75 \times 325)$. We set $N = 10{,}000$, $\rho = 0.1$, and $b = \eta$.

Table 4.2 Dynamics of Algorithm 4.2 for the 3-SAT with matrix $A = (75 \times 325)$

t	$\widehat{\lvert\mathcal{X}_t^*\rvert}$	$\widehat{\lvert\mathcal{X}_t^*\rvert}_{\text{dir}}$	N_t	$N_t^{(s)}$	m_t^*	m_{*t}	$\widehat{c_t}$
1	5.4e+020	0.0	1020	1020	305	292	0.11
2	5.6e+019	0.0	1714	1714	307	297	0.14
3	6.5e+018	0.0	1070	1070	309	301	0.10
4	1.2e+018	0.0	1462	1462	310	304	0.12
5	1.7e+017	0.0	2436	2436	312	306	0.18
6	2.0e+016	0.0	2166	2166	314	308	0.14
7	6.1e+015	0.0	1501	1501	316	310	0.12
8	4.6e+014	0.0	1115	1115	316	312	0.09
9	2.5e+013	0.0	636	636	319	314	0.06
10	5.0e+012	0.0	2213	2213	320	315	0.23
11	9.5e+011	0.0	2674	2674	321	316	0.20
12	1.6e+011	0.0	1969	1969	320	317	0.19
13	2.5e+010	0.0	1962	1962	321	318	0.17
14	3.3e+009	0.0	1775	1775	322	319	0.16
15	3.9e+008	0.0	1350	1350	323	320	0.13
16	3.5e+008	0.0	1437	1437	324	321	0.12
17	3.8e+007	0.0	1270	1270	324	322	0.10
18	2.8e+006	8.0	924	924	325	323	0.08
19	1.4e+005	179.0	537	534	325	324	0.05
20	2341.0	2203.0	196	187	325	325	0.01
21	2341.0	2225.0	10472	2199	325	325	1.00

Table 4.3 Comparison of the Performance of the Adaptive Estimator $\widehat{\lvert\mathcal{X}^*\rvert}$ and Its CAP-RECAP counterpart $\widehat{\lvert\mathcal{X}^*\rvert}_{\text{cap}}$ for the 3-SAT (75×325) model

Run	Iter's	$\widehat{\lvert\mathcal{X}^*\rvert}$	RE of $\widehat{\lvert\mathcal{X}^*\rvert}$	$\widehat{\lvert\mathcal{X}^*\rvert}_{\text{cap}}$	RE of $\widehat{\lvert\mathcal{X}^*\rvert}_{\text{cap}}$	N_1	N_2	R
1	24	2.02E+03	1.03E-02	2.21E+03	6.55E-03	2201	2195	2191
2	24	1.94E+03	2.95E-02	2.20E+03	7.01E-03	2200	2202	2198
3	24	1.59E+03	2.03E-01	2.24E+03	7.86E-03	2234	2235	2232
4	24	2.34E+03	1.70E-01	2.23E+03	2.45E-03	2221	2223	2219
5	24	1.69E+03	1.54E-01	2.20E+03	1.11E-02	2194	2191	2190
6	24	2.38E+03	1.89E-01	2.24E+03	6.96E-03	2230	2230	2225
7	24	1.63E+03	1.86E-01	2.22E+03	6.53E-04	2215	2216	2210
8	24	2.38E+03	1.89E-01	2.23E+03	5.15E-03	2225	2229	2223
9	24	1.97E+03	1.66E-02	2.22E+03	1.55E-03	2217	2219	2213
10	24	2.12E+03	6.03E-02	2.21E+03	3.86E-03	2206	2208	2203
Avg	24	2.01E+03	1.21E-01	2.22E+03	5.31E-03			
Var	0	9.10E+04	6.55E-03	2.04E+02	1.03E-05			

Note that the sample N_1 was obtained as soon as soon as Algorithm 4.6 reaches the final level m, and N_2 was obtained while running it for one more iteration at the same level m. The actual sample sizes N_1 and N_2 were chosen according to the following rule: *sample until Algorithm 4.6 screens out 50% of the samples and then stop*. It follows from Table 4.3 that for model $A = (75 \times 325)$ this corresponds to $N_1 \approx N_2 \approx R \approx 2,200$. It also follows that the relative error of $\widehat{|\mathcal{X}^*|}_{\text{cap}}$ is about 100 times smaller as compared to $\widehat{|\mathcal{X}^*|}$.

4.9.1.4 Random Graphs with Prescribed Degrees

Random graphs with given vertex degrees is a model for real-world complex networks, including the World Wide Web, social networks, and biological networks. The problem is to find a graph $G = G(V, E)$ with n vertices, given the degree sequence $\boldsymbol{d} = (d_1, \ldots, d_n)$ with some nonnegative integers. Following [9], a finite sequence $(d_1, \ldots, d_n)$ of non-negative integers is called *graphical* if there is a labeled simple graph with vertex set $\{1, \ldots, n\}$ in which vertex i has degree d_i. Such a graph is called a realization of the degree sequence $(d_1, \ldots, d_n)$. We are interested in the total number of such realizations for a given degree sequence, hence $\mathcal{X}^*$ denotes the set of all graphs $G = G(V, E)$ with the degree sequence $(d_1, \ldots, d_n)$.

In order to perform this estimation, we transform the problem into a counting problem by considering the complete graph K_n of n vertices, in which each vertex is connected with all other vertices. Thus, the total number of edges in K_n is $m = n(n-1)/2$, labeled $e_1, \ldots, e_m$. The random graph problem with prescribed degrees is translated to the problem of choosing those edges of K_n such that the resulting graph G matches the given sequence $\boldsymbol{d} = (d_1, \ldots, d_n)$. We set $x_i = 1$ when e_i is chosen, and $x_i = 0$, otherwise, $i = 1, \ldots, m$. So that such an assignment $\boldsymbol{x} \in \{0, 1\}^m$ matches the given degree sequence $(d_1, \ldots, d_n)$, it holds necessarily that $\sum_{j=1}^m x_j = \frac{1}{2}\sum_{i=1}^n d_i$, since this is the total number of edges. In other words, the configuration space is

$$\mathcal{X} = \left\{ \boldsymbol{x} \in \{0, 1\}^m : \sum_{j=1}^m x_j = \frac{1}{2}\sum_{i=1}^n d_i \right\}.$$

Let $\boldsymbol{A}$ be the incidence matrix of K_n with entries

$$a_{ij} = \begin{cases} 0 & \text{if } v_i \notin e_j \\ 1 & \text{if } v_i \in e_j. \end{cases}$$

It is easy to see that whenever a configuration $\boldsymbol{x} \in \{0, 1\}^m$ satisfies $\boldsymbol{Ax} = \boldsymbol{d}$, the associated graph has degree sequence $(d_1, \ldots, d_n)$. We conclude that the set $\mathcal{X}^*$ is represented by

$$\mathcal{X}^* = \{\boldsymbol{x} \in \mathcal{X} : \boldsymbol{Ax} = \boldsymbol{d}\}.$$

We first present a small example as illustration. Let $\boldsymbol{d} = (2, 2, 2, 1, 3)$ with $n = 5$, and $m = 10$. After ordering the edges of K_5 lexicographically, the corresponding incidence matrix is given as

$$A = \begin{pmatrix} 1 & 1 & 1 & 1 & 0 & 0 & 0 & 0 & 0 & 0 \\ 1 & 0 & 0 & 0 & 1 & 1 & 1 & 0 & 0 & 0 \\ 0 & 1 & 0 & 0 & 1 & 0 & 0 & 1 & 1 & 0 \\ 0 & 0 & 1 & 0 & 0 & 1 & 0 & 1 & 0 & 1 \\ 0 & 0 & 0 & 1 & 0 & 0 & 1 & 0 & 1 & 1 \end{pmatrix}.$$

It can be seen that $\boldsymbol{x} = (0, 0, 1, 1, 1, 0, 1, 0, 1, 0)'$ and $\boldsymbol{x} = (1, 0, 0, 1, 1, 0, 0, 0, 1, 1)'$ present two solutions of this example.

For the random graph problem we define the performance function $S : \mathcal{X} \to \mathbb{Z}_-$ by

$$S(\boldsymbol{x}) = -\sum_{i=1}^{n} |\deg(v_i) - d_i|,$$

where $\deg(v_i)$ is the degree of vertex i under the configuration $\boldsymbol{x}$. Each configuration that satisfies the degree sequence will have a performance function equal to 0.

The implementation of the Gibbs sampler for this problem is slightly different than for the 3-SAT one, in as much as we keep the number of edges in each realization fixed to $\sum d_i/2$. The algorithm below takes care of this requirement and generates a random $\boldsymbol{x} \in \mathcal{X}$.

Algorithm 4.10 ***A Random Graph with Prescribed Degrees***

- *Input: The prescribed degrees sequence* $(d_1, \ldots, d_n)$.
- *Output: A random* $\boldsymbol{x} \in \mathcal{X}$.

1. *Generate a random permutation of* $1, \ldots, m$.
2. *Choose the first* $\sum d_i/2$ *places in this permutation and deliver a vector* $\boldsymbol{x}$ *having one's in those places.*

The adaptive thresholds in the splitting algorithm are negative, increasing to 0:

$$m_1 \le m_2 \le \cdots \le m_{T-1} \le m_T = 0.$$

The resulting Gibbs sampler (in Step 3 of the splitting algorithm starting with a configuration $\boldsymbol{x} \in \mathcal{X}$ for which $S(\boldsymbol{x}) \ge m_t$) can be written as follows.

Algorithm 4.11 ***Gibbs Algorithm for Random Graph Sampling***

- *Input: A configuration* $\boldsymbol{x} \in \mathcal{X}$ *with* $S(\boldsymbol{x}) \ge m_t$.
- *Output: A uniformly chosen random neighbor with the same property.*

For each edge $x_i = 1$, while keeping all other edges fixed, do:

1. *Remove x_i from $\mathbf{x}$; that is, make $x_i = 0$.*
2. *Check all possible placements for the edge resulting a new vector, denoted by $\bar{\mathbf{x}}$ conditioning on the performance function $S(\bar{\mathbf{x}}) \geq m_t$.*
3. *With uniform probability choose one of the valid realizations.*

We will apply the splitting algorithm to two problems taken from [9].

4.9.1.5 A Small Problem

For this small problem we have the degree sequence

$$\boldsymbol{d} = (5, 6, \underbrace{1, \ldots, 1}_{11 \text{ ones}}).$$

The solution can be obtained analytically and is already given in [9], from which we quote:

> To count the number of labeled graphs with this degree sequence, note that there are $\binom{11}{5} = 462$ such graphs with vertex 1 not joined to vertex 2 by an edge (these graphs look like two separate stars), and there are $\binom{11}{4}\binom{7}{5} = 6930$ such graphs with an edge between vertices 1 and 2 (these look like two joined stars with an isolated edge left over). Thus, the total number of realizations of d is 7392.

As we will see, the adaptive splitting Algorithm 4.2 easily handles the problem. First we present a typical dynamics (Table 4.4).

We ran the algorithm 10 times, independently, using sample size $N = 50{,}000$ and rarity parameter $\rho = 0.5$. The average performance was

$$\widehat{|\mathcal{X}^*|} = 7312.2, \ \text{RE} = 0.0271, \ \text{CPU} = 15.539.$$

Table 4.4 Typical Dynamics of the Splitting Algorithm 4.2 for a small problem using $N = 50{,}000$ and $\rho = 0.5$

t	$\widehat{\lvert\mathcal{X}^*\rvert}$	N_t	$N_t^{(s)}$	m_t^*	m_{*t}	$\widehat{c}_t$
1	4.55E+12	29227	29227	−4	−30	0.5845
2	2.56E+12	28144	28144	−4	−18	0.5629
3	1.09E+12	21227	21227	−6	−16	0.4245
4	3.38E+11	15565	15565	−4	−14	0.3113
5	7.51E+10	11104	11104	−4	−12	0.2221
6	1.11E+10	7408	7408	−2	−10	0.1482
7	1.03E+09	4628	4628	−2	−8	0.0926
8	5.37E+07	2608	2608	−2	−6	0.0522
9	1.26E+06	1175	1175	0	−4	0.0235
10	7223.9	286	280	0	−2	0.0057

4.9.1.6 A Large Problem

A much harder instance (see [9]) is defined by

$$\boldsymbol{d} = (7, 8, 5, 1, 1, 2, 8, 10, 4, 2, 4, 5, 3, 6, 7, 3, 2, 7, 6, 1, 2, 9, 6, 1, 3, 4, 6, 3, 3, 3, 2, 4, 4).$$

In [9] the number of such graphs is estimated to be about 1.533×10^{57}. The average performance with the adaptive splitting Algorithm 4.2, based on 10 experiments, using sample size $N = 100{,}000$ and rarity parameter $\rho = 0.5$, was

$$|\widehat{\mathcal{X}^*}| = 1.59\text{E}{+}57, \text{RE} = 0.0843, \ \text{CPU} = 4253.$$

4.9.1.7 Binary Contingency Tables

Given are two vectors of positive integers $\boldsymbol{r} = (r_1, \ldots, r_m)$ and $\boldsymbol{c} = (c_1, \ldots, c_n)$ such that $r_i \leq n$ for all i, $c_j \leq n$ for all j, and $\sum_{i=1}^m r_i = \sum_{j=1}^n c_j$. A binary contingency table with row sums $\boldsymbol{r}$ and column sums $\boldsymbol{c}$ is a $m \times n$ matrix $\boldsymbol{X}$ of zero-one entries x_{ij} satisfying $\sum_{j=1}^n x_{ij} = r_i$ for every row i and $\sum_{i=1}^m x_{ij} = c_j$ for every column j. The problem is to count all contingency tables.

The extension of the proposed Gibbs sampler for counting the contingency tables follows immediately. We define the configuration space $\mathcal{X} = \mathcal{X}^{(r)} \cup \mathcal{X}^{(c)}$ as the space where all column or row sums are satisfied:

$$\mathcal{X}^{(c)} = \left\{ \boldsymbol{X} \in \{0, 1\}^{m+n} : \sum_{i=1}^{m} x_{ij} = c_j \ \forall j \right\},$$

$$\mathcal{X}^{(r)} = \left\{ \boldsymbol{X} \in \{0, 1\}^{m+n} : \sum_{j=1}^{n} x_{ij} = r_i \ \forall i \right\}.$$

Clearly, sampling uniformly on $\mathcal{X}$ is straightforward. The sample function $S : \mathcal{X} \to \mathbb{Z}_-$ is defined by

$$S(\boldsymbol{X}) = \begin{cases} -\sum_{i=1}^m |\sum_{j=1}^n x_{ij} - r_i|, & \text{for } \boldsymbol{X} \in \mathcal{X}^{(c)}, \\ -\sum_{j=1}^n |\sum_{i=1}^m x_{ij} - c_j|, & \text{for } \boldsymbol{X} \in \mathcal{X}^{(r)}, \end{cases}$$

that is, the difference of the row sums $\sum_{j=1}^n x_{ij}$ with the target r_i if the column sums are right, and vice versa.

The Gibbs sampler is similar to random graphs with prescribed degrees.

Algorithm 4.12 ***Gibbs Algorithm for Random Contingency Tables Sampling***

- *Input: A matrix realization $X \in \mathcal{X}^{(c)}$ with performance value $S(X \geq m_t)$.*
- *Output: A uniformly chosen random neighbor with the same property.*

For each column j and for each 1-entry in this column ($x_{ij} = 1$) do:

1. *Remove this 1, that is, set $x'_{ij} = 0$.*
2. *Check all possible placements for this 1 in the given column j conditioning on the performance function $S(X') \geq m_t$, where X' is the matrix resulting by setting $x'_{ij} = 0$, $x'_{i'j} = 1$ for some $x_{i'j} = 0$, and all other entries remain unchanged.*
3. *Suppose that the set of valid realization is $\mathcal{A} = \{X' | S(X') \geq m_t\}$. This set also contains the original realization X. Then, with probability $\frac{1}{|\mathcal{A}|}$, pick any realization at random and continue with step 1.*

In this way we keep the column sums correct. Similarly, when we started with a matrix configuration with all row sums correct, we execute these steps for each row and swap 1 and 0 per row.

4.9.1.8 Model 1

The data are $m = 12, n = 12$ with row and column sums

$$r = (2, 2, 2, 2, 2, 2, 2, 2, 2, 2, 2, 2),$$
$$c = (2, 2, 2, 2, 2, 2, 2, 2, 2, 2, 2, 2).$$

The true count value is known to be 21,959,547,410,077,200 (reported in [28]). Table 4.5 presents a typical dynamics of the adaptive splitting Algorithm 4.2 for Model 1 using $N = 50{,}000$ and $\rho = 0.5$.

Table 4.5 Typical Dynamics of the Splitting Algorithm 4.2 for model 1 using $N = 50{,}000$ and $\rho = 0.5$

t	$\widehat{\|\mathcal{X}^*\|}$	N_t	$N_t^{(s)}$	m_t^*	m_{*t}	$\widehat{c}_t$
1	4.56E+21	13361	13361	−2	−24	0.6681
2	2.68E+21	11747	11747	−2	−12	0.5874
3	1.10E+21	8234	8234	−2	−10	0.4117
4	2.76E+20	5003	5003	−2	−8	0.2502
5	3.45E+19	2497	2497	0	−6	0.1249
6	1.92E+18	1112	1112	0	−4	0.0556
7	2.08E+16	217	217	0	−2	0.0109

The average performance based on 10 experiments using sample size $N = 50{,}000$ and rarity parameter $\rho = 0.5$ was

$$|\widehat{\mathcal{X}^*}| = 2.17\text{E+16},\ \text{RE} = 0.0521,\ \ \text{CPU} = 4.56.$$

4.9.1.9 Model 2

Consider the problem of counting tables with Darwin's finch data as marginal sums given in Chen et al. [28]. The data are $m = 12, n = 17$ with row and columns sums

$$\boldsymbol{r} = (14, 13, 14, 10, 12, 2, 10, 1, 10, 11, 6, 2),$$
$$\boldsymbol{c} = (3, 3, 10, 9, 9, 7, 8, 9, 7, 8, 2, 9, 3, 6, 8, 2, 2).$$

The true count value in [28] is $67, 149, 106, 137, 567, 600$. The average performance based on 10 independent experiments using sample size $N = 200{,}000$ and rarity parameter $\rho = 0.5$ was

$$|\widehat{\mathcal{X}^*}| = 6.71\text{E+16},\ \text{RE} = 0.0785,\ \ \text{CPU} = 259.95.$$

4.9.1.10 Vertex Coloring

Vertex coloring is a problem of coloring the vertices of a graph $G = G(V, E)$ consisting of $n = |V|$ nodes and $m = |E|$ edges, such that neighboring vertices have different colors. The set of colors has a size q. For more details on vertex coloring see Section A.1.1.

Algorithm 4.13 ***Gibbs Algorithm for Vertex Coloring***

- *Input: A graph $G(V, E)$ with proper coloring.*
- *Output: A uniformly chosen random neighbor with the same property.*

For each vertex $v \in V$ do:

1. *Choose a color for v according to the uniform distribution over the set of colors that are not assigned at any of its neighbors.*
2. *Accept the new color with probability $\frac{1}{2}$ and leave all other vertices unchanged.*

We consider here two coloring models: a small one and one of moderate size.

4.9.1.11 First Model: $n = 20$ Vertices, $m = 62$ Edges, $q = 21$ Colors

The graph G has the following $\boldsymbol{A} = 20 \times 20$ adjacency matrix.

$$
\begin{pmatrix}
0 & 1 & 1 & 0 & 0 & 0 & 0 & 1 & 1 & 1 & 0 & 0 & 0 & 1 & 0 & 0 & 0 & 0 & 1 & 0 \\
1 & 0 & 1 & 1 & 1 & 1 & 0 & 0 & 0 & 0 & 0 & 0 & 0 & 1 & 0 & 0 & 0 & 1 & 1 & 0 \\
1 & 1 & 0 & 1 & 1 & 1 & 0 & 1 & 0 & 1 & 0 & 0 & 0 & 0 & 0 & 1 & 0 & 0 & 1 & 0 \\
0 & 1 & 1 & 0 & 0 & 1 & 1 & 0 & 1 & 0 & 0 & 0 & 0 & 0 & 0 & 0 & 0 & 0 & 0 & 0 \\
0 & 1 & 1 & 0 & 0 & 0 & 0 & 0 & 0 & 0 & 0 & 0 & 0 & 0 & 1 & 0 & 0 & 1 & 0 & 0 \\
0 & 1 & 1 & 1 & 0 & 0 & 0 & 1 & 0 & 0 & 0 & 1 & 0 & 0 & 1 & 0 & 0 & 0 & 1 & 0 \\
0 & 0 & 0 & 1 & 0 & 0 & 0 & 0 & 0 & 0 & 0 & 0 & 0 & 1 & 0 & 0 & 0 & 0 & 1 & 0 \\
1 & 0 & 1 & 0 & 0 & 1 & 0 & 0 & 1 & 0 & 1 & 0 & 0 & 0 & 0 & 0 & 0 & 0 & 0 & 0 \\
1 & 0 & 0 & 1 & 0 & 0 & 0 & 1 & 0 & 0 & 0 & 1 & 0 & 0 & 0 & 0 & 0 & 1 & 0 & 0 \\
1 & 0 & 1 & 0 & 0 & 0 & 0 & 0 & 0 & 0 & 0 & 1 & 0 & 0 & 0 & 1 & 1 & 0 & 0 & 0 \\
0 & 0 & 0 & 0 & 0 & 0 & 0 & 1 & 0 & 0 & 0 & 0 & 0 & 1 & 1 & 0 & 1 & 1 & 1 & 0 \\
0 & 0 & 0 & 0 & 0 & 1 & 0 & 0 & 1 & 1 & 0 & 0 & 1 & 0 & 0 & 1 & 0 & 0 & 1 & 1 \\
0 & 0 & 0 & 0 & 0 & 0 & 0 & 0 & 0 & 0 & 0 & 1 & 0 & 0 & 0 & 0 & 1 & 1 & 0 & 1 \\
1 & 1 & 0 & 0 & 0 & 0 & 1 & 0 & 0 & 0 & 1 & 0 & 0 & 0 & 1 & 0 & 1 & 1 & 0 & 0 \\
0 & 0 & 0 & 0 & 1 & 1 & 0 & 0 & 0 & 0 & 1 & 0 & 0 & 1 & 0 & 0 & 1 & 0 & 1 & 0 \\
0 & 0 & 1 & 0 & 0 & 0 & 0 & 0 & 0 & 1 & 0 & 1 & 0 & 0 & 0 & 0 & 0 & 1 & 0 & 1 \\
0 & 0 & 0 & 0 & 0 & 0 & 0 & 0 & 0 & 1 & 1 & 0 & 1 & 1 & 1 & 0 & 0 & 0 & 1 & 1 \\
0 & 1 & 0 & 0 & 1 & 0 & 0 & 0 & 1 & 0 & 1 & 0 & 1 & 1 & 0 & 1 & 0 & 0 & 0 & 1 \\
1 & 1 & 1 & 0 & 0 & 1 & 1 & 0 & 0 & 0 & 1 & 1 & 0 & 0 & 1 & 0 & 1 & 0 & 0 & 1 \\
0 & 0 & 0 & 0 & 0 & 0 & 0 & 0 & 0 & 0 & 0 & 1 & 1 & 0 & 0 & 1 & 1 & 1 & 1 & 0
\end{pmatrix}
$$

The performance is based on 10 independent experiments using sample size $N =$ 1,000 and rarity parameter $\rho = 0.05$, was

$$\widehat{|\mathcal{X}^*|} = 1.25\text{E+}25,\ \text{RE} = 0.09,\ \text{CPU} = 0.811.$$

Note that the splitting procedure starts with the set $\mathcal{X} = \mathcal{X}_0$ consisting of all feasible colorings of the graph $G_0 = (V, \emptyset)$ (see Section A.1.1). Thus, $|\mathcal{X}| = q^n$. We obtain, in this case, that the probability equals

$$\ell = \frac{|\mathcal{X}^*|}{|\mathcal{X}|} \approx 0.045.$$

Because ℓ is not a rareevent, the crude Monte Carlo method can also be used here. The performance of the CMC estimator based on 10 independent runs with $N = 1{,}000$ and $\rho = 0.05$ was

$$\widehat{|\mathcal{X}^*|}_{\text{CMC}} = 1.19\text{E+}25,\ \text{RE} = 0.3,\ \text{CPU} = 1.033.$$

It follows the RE of the CMC is about 3 times larger than its counterpart, the splitting Algorithm 4.2.

4.9.1.12 Second Model: $n = 40$ Vertices, $m = 162$ Edges, $q = 4$

We next consider a larger and sparser coloring model, one with $n = 40$ vertices, $m = 162$ edges, and $q = 4$.

Table 4.6 Dynamics of Algorithm 4.2 for 4-Coloring graph with $n = 40$ nodes

t	$\widehat{\lvert\mathcal{X}_t^*\rvert}$	$\widehat{\lvert\mathcal{X}_t^*\rvert}_{\mathrm{dir}}$	$N_t^{(e)}$	$N_t^{(s)}$	m_t^*	m_{*t}	$\widehat{c}_t$
1	7.97E+22	0	6591	6591	−40	−64	6.59E-02
2	6.94E+21	0	8704	8704	−32	−54	8.70E-02
3	4.43E+20	0	6384	6384	−28	−46	6.38E-02
4	3.23E+19	0	7299	7299	−26	−40	7.30E-02
5	4.32E+18	0	13363	13363	−22	−36	1.34E-01
6	4.49E+17	0	10397	10397	−20	−32	1.04E-01
7	3.52E+16	0	7835	7835	−16	−28	7.84E-02
8	2.13E+15	0	6045	6045	−14	−24	6.05E-02
9	4.58E+14	0	21532	21531	−12	−22	2.15E-01
10	8.96E+13	0	19569	19569	−12	−20	1.96E-01
11	1.62E+13	0	18092	18092	−10	−18	1.81E-01
12	2.66E+12	0	16378	16378	−8	−16	1.64E-01
13	3.80E+11	0	14294	14291	−6	−14	1.43E-01
14	4.74E+10	0	12479	12472	−6	−12	1.25E-01
15	5.06E+09	0	10685	10675	−4	−10	1.07E-01
16	4.47E+08	0	8842	8825	−2	−8	8.84E-02
17	3.15E+07	0	7047	7003	−2	−6	7.05E-02
18	1.69E+06	1	5372	5261	0	−4	5.37E-02
19	5.99E+04	85	3537	3241	0	−2	3.54E-02
20	1.38E+03	1316	100000	1316	0	0	1.00E+00

For this model, we applied the adaptive splitting Algorithm 4.2 with sample size $N = 100{,}000$, rarity parameter $\rho_t \equiv 0.05$, and burn-in parameter $b_t \equiv 1$. The direct algorithm was implemented immediately after the last iteration of the adaptive splitting, using again $N = 100{,}000$. Table 4.6 presents the dynamics for a single run of the splitting Algorithm 4.2.

The average performance based on 10 independent experiments was

$$\widehat{|\mathcal{X}^*|} = 1400.7, \ \mathrm{RE} = 0.09, \ \mathrm{CPU} = 258.144.$$

The direct splitting algorithm results in $\widehat{|\mathcal{X}^*|}_{\mathrm{dir}} = 1314$ with $\mathrm{RE} = 0.021$.

4.9.1.13 Permanent

The permanent of a general $n \times n$ matrix $\boldsymbol{A} = (a_{ij})$ is defined as

$$\mathrm{perm}(\boldsymbol{A}) = |\mathcal{X}^*| = \sum_{\boldsymbol{x} \in \mathcal{X}} \prod_{i=1}^{n} a_{ix_i}, \tag{4.27}$$

where $\mathcal{X}$ is the set of all permutations $\boldsymbol{x} = (x_1, \ldots, x_n)$ of $(1, \ldots, n)$.

EXAMPLE 4.2

Let $\boldsymbol{A}$ be a 3×3 matrix given as

$$\boldsymbol{A} = \begin{pmatrix} 1 & 1 & 1 \\ 1 & 1 & 0 \\ 0 & 1 & 1 \end{pmatrix}. \tag{4.28}$$

Note that, in contrast to the Hamiltonian cycle problem, the diagonal elements of the matrix $\boldsymbol{A}$ need not to be zeros. The $3! = 6$ permutations with the corresponding product values $\prod_{i=1}^{3} a_{ix_i}$ are given in Table 4.7. The permanent is $|\mathcal{X}^*| = 3$. The values of the performance function $S(\boldsymbol{x})$ with all six possible permutation vectors $\boldsymbol{x}$ are given in Table 4.8. □

To apply the Gibbs sampler for the permanent problem, we adopt the concept of "neighboring" elements (see, for example, Chapter 10 of Ross [99]) and in particular the concept of 2-opt heuristic [108]. More precisely, given a point (tour) $\boldsymbol{x}$ of length m_t generated by the pdf $g_t = g(\boldsymbol{x}, m_t)$, the conditional Gibbs sampling updates the existing tour $\boldsymbol{x}$ to $\tilde{\boldsymbol{x}}$, where $\tilde{\boldsymbol{x}}$ is generated from $g(\tilde{\boldsymbol{x}}, m_t)$ and where $\tilde{\boldsymbol{x}}$ is the same as $\boldsymbol{x}$ with the exception that the points x_i and x_j in $\tilde{\boldsymbol{x}}$ are interchanged. We accept the tour $\tilde{\boldsymbol{x}}$ if $S(\tilde{\boldsymbol{x}}) \geq m_t$, otherwise, we leave the tour $\boldsymbol{x}$ the same. If $\tilde{\boldsymbol{x}}$ is accepted, we update the cost function $S(\boldsymbol{x})$ (for $j > i$) accordingly. We found

Table 4.7 Six permutations and their product values of the matrix $\boldsymbol{A}$ in (4.28)

$\boldsymbol{x}$			$\prod_{i=1}^{3} a_{ix_i}$	
1	2	3	$a_{11}\, a_{22}\, a_{33}$	$= 1$
1	3	2	$a_{11}\, a_{23}\, a_{32}$	$= 0$
2	1	3	$a_{12}\, a_{21}\, a_{33}$	$= 1$
2	3	1	$a_{12}\, a_{23}\, a_{31}$	$= 0$
3	1	2	$a_{13}\, a_{21}\, a_{32}$	$= 1$
3	2	1	$a_{13}\, a_{22}\, a_{31}$	$= 0$

Table 4.8 The $S(\boldsymbol{x})$ values corresponding to the permutations for the matrix $\boldsymbol{A}$ in (4.28)

$S(\boldsymbol{x})$	
$a_{11} + a_{22} + a_{33}$	$= 3$
$a_{11} + a_{23} + a_{32}$	$= 2$
$a_{12} + a_{21} + a_{33}$	$= 3$
$a_{12} + a_{23} + a_{31}$	$= 1$
$a_{13} + a_{21} + a_{32}$	$= 3$
$a_{13} + a_{22} + a_{31}$	$= 2$

empirically that, in order for the tours $\boldsymbol{X}$ to be distributed approximately uniformly (at each subspace $\mathcal{X}_t$), we used $n(n-1)/2$ trials.

The Gibbs sampler may be summarized as follows:

Algorithm 4.14 ***Gibbs Algorithm for the Permanent***

- *Input: Given a permutation $\boldsymbol{x} = \{x_1, \ldots, x_n\}$ and $S(\boldsymbol{x}) \geq m_t$.*
- *Output: A uniformly chosen random neighbor with the same property.*

For each i and j $i \leq j$ do:

1. $\tilde{\boldsymbol{x}} = (x_1, \ldots, x_{i-1}, x_j, x_{i+1}, \ldots, x_{j-1}, x_i, x_{j+1}, \ldots, x_n)$;
2. *If $S(\tilde{\boldsymbol{x}}) \geq m_t$, set $\boldsymbol{x} = \tilde{\boldsymbol{x}}$.*

We applied the adaptive splitting Algorithm 4.2 to a problem with an instance matrix $\boldsymbol{A} = 30 \times 30$ matrix. We took a sample size $N = 100{,}000$ and set the rarity parameter $\rho_t \equiv 0.1$ and burn-in parameter $b_t \equiv 1$. The average performance based on 10 experiments was

$$\widehat{|\mathcal{X}^*|} = 261.94, \quad \mathrm{RE} = 0.022, \quad \mathrm{CPU} = 116.48.$$

The direct estimator was implemented immediately after the last iteration of the adaptive splitting, using again $N = 100{,}000$. The average performance of the direct algorithm was $\widehat{|\mathcal{X}^*|}_{\mathrm{dir}} = 266$ with $\mathrm{RE} = 0$.

4.9.1.14 Hamiltonian Cycles

We solve the Hamiltonian cycles problem by applying the 2-opt heuristic used for the permanent. In counting Hamiltonian cycles we simulated an associated permanent problem by making use of the following observations:

- The set of Hamiltonian cycles of length n represents a subset of the associated permanent set of trajectories.
- The latter set is formed from cycles of length $\leq n$.

It follows that, in order to count the number of Hamiltonian cycles for a given matrix $\boldsymbol{A}$, one can use the following simple procedure:

1. Run Algorithm 4.2 and calculate the estimator of $|\mathcal{X}^*|$ of the associated permanent using the product formula (4.9). Denote such a permanent estimator by $\widehat{|\mathcal{X}^*|}_p$.
2. Proceed with one more iteration of Algorithm 4.2 and calculate the ratio of the number of screened Hamiltonian elite cycles (cycles of length n) to that of screened elite samples (samples of length $\leq n$) in the permanent. Denote the ratio by ζ.

3. Deliver $\widehat{|\mathcal{X}^*|}_H = \zeta \widehat{|\mathcal{X}^*|}_p$ as an estimator of the number Hamiltonian cycles $|\mathcal{X}^*|$.

Below we present simulation results for two 30×30 models. We choose $N = 100{,}000$, $\rho_t \equiv 0.1$ and $b_t = \eta_t$. The results are based on the averages of 10 experiments. The adaptive splitting algorithm results in

$$\widehat{|\mathcal{X}^*|} = 12.186, \ \text{RE} = 0.069, \ \text{CPU} = 117.05;$$

while the direct algorithm results in

$$\widehat{|\mathcal{X}^*|}_{\text{dir}} = 12, \ \text{RE} = 0.$$

4.9.1.15 Volume of a Polytope

Here we present numerical results with the splitting Algorithm 4.2, on the volume of a polytope defined as $\boldsymbol{Ax} \le \boldsymbol{b}$. We assume in addition that all components $x_k \in (0, 1)$, $k = 1, \ldots, n$. By doing so the polytope $\boldsymbol{Ax} \le \boldsymbol{b}$ will be inside the unit n-dimensional cube and therefore its volume is very small (rare-event probability) relative to that of the unit cube. In general, however, before implementing the Gibbs sampler, one must find the minimal n-dimensional box, which contains the polytope, that is, for each i, $i = 1, \ldots, n$, one must find the values (c_i, d_i) of the minimal box satisfying $c_i \le x_i \le d_i$, $i = 1, \ldots, n$. This can be done by solving for each i, $i = 1, \ldots, n$ the following two linear programs:

$$\min \ x_i \tag{4.29}$$
$$\text{s.t. } \boldsymbol{Ax} \le \boldsymbol{b} \tag{4.30}$$

and

$$\max \ x_i \tag{4.31}$$
$$\text{s.t. } \boldsymbol{Ax} \le \boldsymbol{b}. \tag{4.32}$$

We consider two models.

4.9.1.16 First Model

$$\boldsymbol{A} = \begin{pmatrix} 1 & 0 & 1 & 0 & 0 & 0 & 0 & 0 & 0 & 0 \\ 1 & 0 & 0 & 0 & 0 & 0 & 1 & 0 & 0 & 0 \\ 0 & 1 & 0 & 0 & 0 & 1 & 0 & 0 & 0 & 0 \\ 0 & 0 & 1 & 0 & 0 & 1 & 0 & 0 & 0 & 1 \\ 0 & 1 & 1 & 0 & 0 & 0 & 0 & 0 & 0 & 0 \\ 1 & 0 & 0 & 0 & 0 & 1 & 0 & 1 & 0 & 0 \\ 0 & 1 & 0 & 0 & 0 & 0 & 0 & 0 & 0 & 1 \\ 0 & 0 & 0 & 0 & 1 & 0 & 0 & 0 & 1 & 0 \\ 0 & 0 & 0 & 0 & 1 & 0 & 1 & 0 & 0 & 0 \\ 1 & 0 & 0 & 0 & 0 & 0 & 1 & 1 & 0 & 0 \end{pmatrix} \quad \text{and } \boldsymbol{b} = \begin{pmatrix} 0.7 \\ 0.7 \\ 0.7 \\ 1.4 \\ 0.7 \\ 1.4 \\ 0.7 \\ 0.7 \\ 0.7 \\ 1.4. \end{pmatrix} \tag{4.33}$$

Table 4.9 Dynamics of single run of splitting Algorithm 4.2

t	$N_t^{(e)}$	$N_t^{(s)}$	m_t^*	m_{*t}	$\widehat{c}_t$
1	2377	2377	10	5	0.2377
2	3770	3770	10	7	0.2643
3	1485	1485	10	9	0.1313
4	2654	2654	10	10	0.2234
5	10616	10616	10	10	1.0000

Table 4.9 presents the dynamics of a single run of the splitting Algorithm 4.2 with $N = 10{,}000$ and $\rho_t \equiv 0.3$. The results are self-explanatory.

In this model, we also executed the crude Monte Carlo, for which we took a sample $N = 10^6$. The results of 10 experiments are

$$\widehat{|\mathcal{X}^*|} = 0.0018, \;\; \text{RE} = 0.03, \;\; \text{CPU} = 12.$$
$$\widehat{|\mathcal{X}^*|}_{\text{CMC}} = 0.0018, \;\; \text{RE} = 0.05, \;\; \text{CPU} = 18.$$

4.9.1.17 Second Model

This problem involves a (61×30) matrix, whose main part, (60×30), comprises the first 60 column vectors of matrix $\boldsymbol{A}$ with 0-1 entries and the rightmost column corresponds to the vector $\boldsymbol{b}$. Running the splitting Algorithm 4.2 for 10 independent runs, each with $N = 10{,}000$ and $\rho_t \equiv 0.3$, we obtained average performance

$$\widehat{|\mathcal{X}^*|} = 1.1387\text{E-}29, \text{RE} = 0.02, \;\; \text{CPU} = 56.$$

4.9.2 Combinatorial Optimization

4.9.2.1 Traveling Salesman Problem

We now present some numerical results obtained with splitting Algorithm 4.2 while using the 2-opt heuristics. The models are taken from http://www.iwr.uni-heidelberg.de/groups/comopt/software/TSPLIB95/atsp.

All our results are based on 10 independent runs with the random Gibbs sampler. Below, we consider the `ftv33` model with $n = 34$ nodes and the best-known solution, equal to 1286. We used in our simulation studies $b_t = \eta_t$ and $b_t = 1$. We found that the accuracy with $b_t = \eta_t$ is higher than for $b_t = 1$, but the CPU time in the latter increased by about 20%. We also found that the accuracy of the splitting Algorithm 4.2 is quite sensitive to N, ρ, and b. Below are some results:

- $N = 10{,}000$, $\rho_t \equiv 0.1$, and $b_t = \eta_t$. The results are

  $$1286, \; 1329, \; 1286, \; 1302, \; 1311, \; 1323, \; 1286, \; 1311, \; 1302, \; 1311.$$

  The CPU was 25 sec for a run.

- $N = 10{,}000$, $\rho_t \equiv 0.3$, and $b_t = \eta_t$. In this case, the 2-opt heuristic results in the best-known solution is 1286 in all 10 runs.
- $N = 10{,}000$, $\rho_t \equiv 0.3$, and $b_t \equiv 1$. The results are

  1286, 1286, 1302, 1286, 1302, 1286, 1286, 1286, 1286, 1286.

- $N = 2{,}000$, $\rho_t \equiv 0.1$, and $b_t = \eta_t$. The results are

  1302, 1324, 1363, 1329, 1328, 1286, 1302, 1302, 1347, 1324.

- $N = 2{,}000$, $\rho_t \equiv 0.1$, and $b_t \equiv 1$. The results are

  1388, 1320, 1366, 1345, 1366, 1397, 1367, 1325, 1347, 1370.

  (This is the worst case, with relative error about 7%.)

4.9.2.2 Knapsack Problem

Consider the `Sento1.dat` knapsack problem given in http://people.brunel.ac.uk/mastjjb/jeb/orlib/files/mknap2.txt. The problem has 30 constraints and 60 variables. We ran the splitting Algorithm 4.4 for 10 independent runs with $\rho = 0.1$, $N = 1000$, $b = 5$, and we found that it discovers the best-known solution. We also ran it for various integer problems and always found that the results coincided with the best-known solution.

4.9.3 Reliability Models

To estimate the system unreliability parameter $\bar{r}$, we use in all our numerical experiments the exponential pdf $f(\boldsymbol{x}, \boldsymbol{\lambda}) \propto \exp(-\sum_i \lambda_i x_i)$ with the rates $\lambda_i = -\ln(1 - p_i)$, where $\boldsymbol{p} = (p_1, \ldots, p_m)$ is the underlying Bernoulli vector.

Our numerical results are given for a dodecahedron graph in Figure 4.11 and a lattice graph. Note that an $m_1 \times m_2$ graph has m_1 rows containing m_2 nodes each, arranged in a rectangular lattice. Each node is connected to its neighbors on the left, right, above, and below, where they exist. As an illustration, a 5×8 lattice graph is presented in Figure 4.12. In all our numerical experiments, we take the terminals to be the upper-left and the lower-right nodes of the corresponding lattice graph.

To estimate the unreliability parameter $\bar{r}$ we run both models for 10 independent scenarios.

4.9.3.1 Dodecahedron Graph with Terminal Nodes $T = \{1, 4, 7, 10, 13, 16, 20\}$

We considered the following two cases:

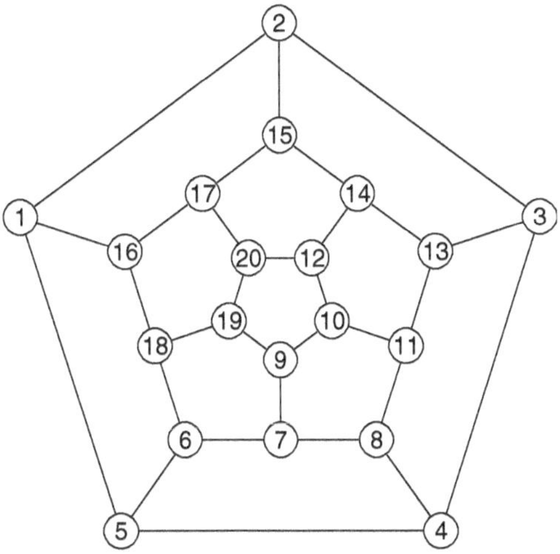

Figure 4.11 The dodecahedron graph.

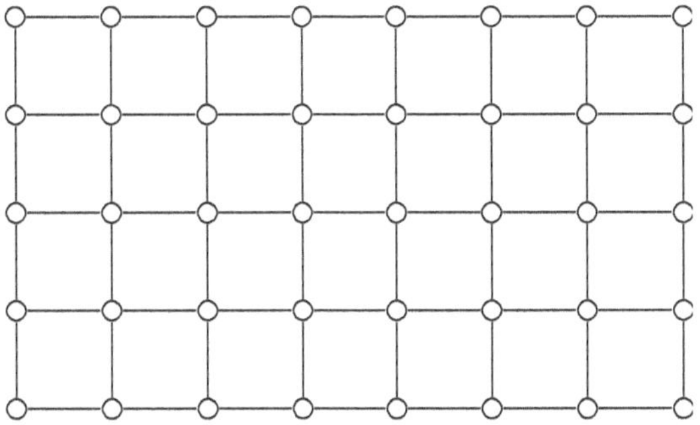

Figure 4.12 A 5×8 lattice graph, with 40 nodes and 67 edges.

Case 1 with $q = 0.01$. We set $N = 10{,}000$ and $\rho = 0.5$. We obtained the following 10 estimators of $\bar{r}$:

6.89E-06, 7.11E-06, 7.07E-06, 7.72E-06, 7.16E-06,
7.30E-06, 7.38E-06, 6.94E-06, 7.70E-06, 7.59E-06.

The average CPU time was 7.04 seconds and the relative error was 3.96%.

Case 2 with $q = 0.000001$. We set $N = 10{,}000$ and $\rho = 0.5$. We obtained the following 10 estimators of $\bar{r}$:

7.41E-18, 7.89E-18, 7.98E-18, 7.12E-18, 6.82E-18,
7.98E-18, 7.29E-18, 7.26E-18, 7.67E-18 7.34E-18.

The average CPU time was 23.2 seconds and the relative error was about 5%.

4.9.3.2 Lattice 50 x 50 Graph with $T = \{0, 2500\}$

We considered the following two cases:

Case 1 with $q = 0.01$. We set $N = 1{,}000$ and $\rho = 0.5$. We obtained the following 10 estimators of $\bar{r}$:

2.09E-04, 1.75E-04, 2.25E-04, 2.22E-04, 1.97E-04,

1.91E-04, 2.30E-04, 2.13E-04, 1.88E-04, 2.24E-04.

The average CPU time was 1382 seconds and the relative error was 8.5%.

Case 2 with $q = 0.000001$, $N = 10{,}000$, and $\rho = 0.5$. We obtained the following 10 estimators of $\bar{r}$:

2.07E-12, 2.14E-12, 2.01E-12, 1.89E-12, 1.74E-12,

1.93E-12, 2.01E-12, 1.96E-12, 2.20E-12, 2.03E-12.

The average CPU time was $1.84E+04$ seconds and the relative error was 6.17%.

4.10 APPENDIX: GIBBS SAMPLER

In this appendix, we show how to sample from a given joint pdf $g(x_1, \ldots, x_n)$ using the Gibbs sampler, where instead of proceeding with $g(x_1, \ldots, x_n)$ directly, which might be very difficult, one samples from one-dimensional conditional pdfs $g(x_i | X_1, \ldots, X_{i-1}, X_{i+1}, \ldots, X_n)$, $i = 1, \ldots, n$, which is typically much simpler.

Two basic versions of the Gibbs sampler [108] are available: *systematic* and *random*. In the former, the components of the vector $\boldsymbol{X} = (X_1, \ldots, X_n)$ are updated in a fixed, say increasing, order, while in the latter they are chosen randomly according to a discrete uniform n-point pdf. Below we present the systematic version, where for a given vector $\boldsymbol{X} = (X_1, \ldots, X_n) \sim g(\boldsymbol{x})$, one generates a new vector $\widetilde{\boldsymbol{X}} = (\widetilde{X}_1, \ldots, \widetilde{X}_n)$ with the same distribution $\sim g(\boldsymbol{x})$ as follows:

Algorithm 4.15 *Systematic Gibbs Sampler*

1. *Draw $\widetilde{X}_1$ from the conditional pdf $g(x_1 | X_2, \ldots, X_n)$.*
2. *Draw $\widetilde{X}_i$ from the conditional pdf $g(x_i | \widetilde{X}_1, \ldots, \widetilde{X}_{i-1}, X_{i+1}, \ldots, X_n)$, $i = 2, \ldots, n-1$.*
3. *Draw $\widetilde{X}_n$ from the conditional pdf $g(x_n | \widetilde{X}_1, \ldots, \widetilde{X}_{n-1})$.*

Iterating with Algorithm 4.15, the Gibbs sampler generates (subject to some mild conditions, see [108]), a sample distributed $g(x_1, \ldots, x_n)$.

We denote for convenience each conditional pdf $g(x_i|\widetilde{X}_1, \ldots, \widetilde{X}_{i-1}, X_{i+1}, \ldots, X_n)$ by $g(x_i|\boldsymbol{x}_{-i})$, where $|\boldsymbol{x}_{-i}$ denotes conditioning on all random variables except the i-th component.

We next present a random Gibbs sampler taken from (Ross, 2006) for estimating each $c_t = \mathbb{E}_{g^*_{t-1}}[I_{\{S(\boldsymbol{X}) \geq m_{t-1}\}}]$, $t = 0, 1, \ldots, T$ according to (4.11), that is,

$$\widehat{c}_t = \frac{1}{N} \sum_{i=1}^{N} I_{\{S(\boldsymbol{X}_i) \geq m_{t-1}\}} = \frac{N_t^{(e)}}{N}.$$

Algorithm 4.16 ***Ross' Acceptance-Rejection Algorithm for Estimating c_t***

1. *Set $N_t^{(e)} = N$=0.*
2. *Choose a vector $\boldsymbol{x}$ such that $S(\boldsymbol{x}) \geq m_{t-1}$.*
3. *Generate a random number $U \sim \mathsf{U}(0, 1)$ and set I=$\lceil nU \rceil + 1$.*
4. *If $I = k$, generate Y_k from the conditional one-dimensional distribution $g(x_k|\boldsymbol{x}_{-k})$ (see Algorithm 4.15).*
5. *If $S(\widetilde{X}_1, \ldots, \widetilde{X}_{k-1}, Y_k, X_{k+1}, \ldots, X_n) < m_{t-1}$, return to 4.*
6. *Set $N = N + 1$ and $Y_k = \widetilde{X}_k$.*
7. *If $S(\boldsymbol{x}) \geq m_t$, then $N_t^{(e)} = N_t^{(e)} + 1$.*
8. *Go to 3.*
9. *Estimate c_t as $\widehat{c}_t = \frac{N_t^{(e)}}{N}$.*

For many rare-event and counting problems, generation from the conditional pdf $g(x_i|\boldsymbol{x}_{-i})$ can be achieved directly, that is, skipping Step 5. This should obviously speed things up.

Chapter 5

Stochastic Enumeration Method

5.1 INTRODUCTION

In this chapter, we introduce a new generic *sequential importance sampling* (SIS) algorithm, called *stochastic enumeration* (SE) for counting #P problems. SE represents a natural generalization of *one-step-look-ahead* (OSLA) algorithms. We briefly introduce these concepts here and defer explanation of the algorithms in detail to the subsequent sections.

Consider a simple walk in the integer lattice $\mathbb{Z}^2$. It starts at the origin (0,0) and it repeatedly takes unit steps in any of the four directions, North (N), South (S), East (E), or West (W). For instance, an 8-step walk could be

$$(0,0) \xrightarrow{(E)} (1,0) \xrightarrow{(E)} (2,0) \xrightarrow{(S)} (2,-1) \xrightarrow{(W)} (1,-1) \xrightarrow{(N)} (1,0) \xrightarrow{(N)} (1,1) \xrightarrow{(N)} (1,2) \xrightarrow{(W)} (0,1).$$

We impose the condition that the walk may not revisit a point that it has previously visited. These walks are called *self-avoiding walks* (SAW). SAWs are often used to model the real-life behavior of chain-like entities such as polymers, whose physical volume prohibits multiple occupation of the same spatial point. They also play a central role in modeling of the topological and knot-theoretic behavior of molecules such as proteins.

Clearly, the 8-step walk above is not a SAW inasmuch as it revisits point $(1,0)$. Figure 5.1 presents an example of a 130-step SAW. Counting the number of SAWs of length n is a basic problem in computational science for which no exact formula is currently known. The problem is conjectured to be complete in $\#P_1$, that is, the version of #P in which the input consists of binary symbols, see [78, 120]. Though many approximating methods exist, the most advanced algorithms are the pivot ones. They can handle SAWs of size 10^7, see [30, 82]. For a good recent survey, see [121].

Fast Sequential Monte Carlo Methods for Counting and Optimization, First Edition.
Reuven Y. Rubinstein, Ad Ridder, and Radislav Vaisman.

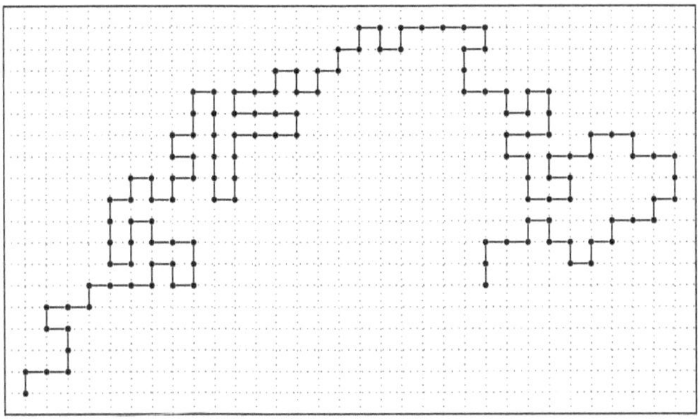

Figure 5.1 SAW of length 130.

We consider randomized algorithms for counting SAWs of length at least some given n, say $n = 500$. A natural, simple algorithm goes as follows: it chooses at the current end point of the walk randomly one of the feasible neighbors and moves to it. This is done until either a path of length of n is obtained or there is no available neighbor. In the latter case, we say that the walk is trapped, or that the algorithm gets stuck. Then, the algorithm returns a zero counting value. Otherwise it returns the product of the consecutive numbers of available neighbors at each step. This is repeated M times independently; the average is taken as an estimator of the counting problem. Typical features of this algorithm are that

- it runs a single path, or trajectory;
- it is sequential, because it moves iteratively from point to point;
- it is randomized, because the next point is chosen at random (if possible);
- it is OSLA, because in each iteration it decides how to continue by considering only its *nearest* feasible points.

Note that any feasible SAW $\boldsymbol{x}$ of length n has a positive probability $g(\boldsymbol{x})$ of being constructed by this simple algorithm. The main two drawbacks of the algorithm are that

- the pdf $g(\boldsymbol{x})$ is nonuniform on the space of all feasible SAWs of length n, which causes high variance of the associated estimator;
- it gets stuck easily for large n, cf. [79].

To overcome the first drawback, we suggest in Section 5.2.2 to run multiple trajectories in parallel. This will typically reduce the variance substantially. To eliminate the second drawback, we ask an oracle at each iteration how to continue in order that the walk not be trapped too early. That is, the oracle can foresee which feasible nearest neighbors lead to a trap and which do not. As an example, consider constructing a SAW of length $n = 20$, then Figure 5.2 shows how the OSLA algorithm can lead to a trap after 15 iterations.

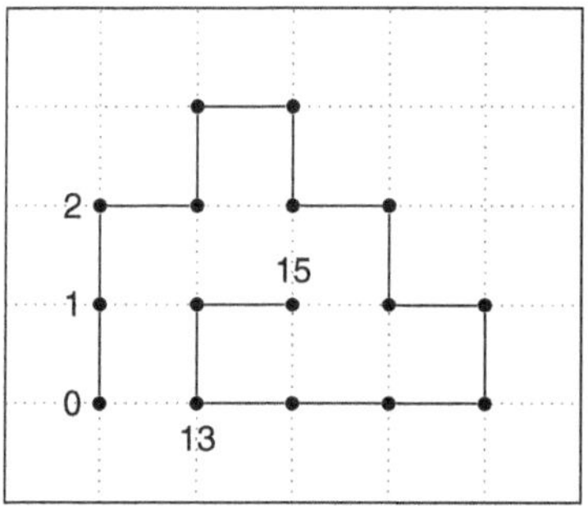

Figure 5.2 SAW trapped after 15 iterations.

However, an oracle would, say, after 13 iterations continue to the South rather than to the North. We call such oracle-based algorithms *n-step-look-ahead* (nSLA), because instances are generated iteratively while having information how to continue without being trapped before the last iteration or step. Thus, it has the advantage of not losing trajectories. Notice that in our example this nSLA algorithm still runs single trajectories, resulting again in high variances of its associated estimator, see Section 5.2.1. Next, we propose to run multiple trajectories in combination with the nSLA oracle. We call this method *stochastic enumeration* (SE), and we will provide the details and analysis in Section 5.3.

Summarizing, the features of SE are

- It runs multiple trajectories in parallel, instead of a single one; by doing so, the samples from $g(\boldsymbol{x})$ become close to uniform.
- It employs n-step-look-ahead algorithms (oracles); this is done to overcome the generation of zero outcomes of the estimator.

As a side effect, SE reduces a difficult counting problem to a set of simple ones, applying at each step an oracle. Furthermore, we note that, in contrast to the conventional splitting algorithm of Chapter 4, there is less randomness involved in SE. As a result, it is typically faster than splitting.

A main concern of using oracles deals with its computational complexity. At each iteration, we apply a decision-making oracle because we are only interested in which direction we can continue generating the path without being trapped somewhere later. Thus, we propose directions and wait for yes-no decisions of the oracle. Hence, for problems for which polynomial time decision-making algorithms exist, SE will be applicable and runs fast.

It is interesting to note that there are many problems in computational combinatorics for which the decision-making is easy (polynomial) but the counting is hard (#P-complete). For instance,

1. How many different variable assignments will satisfy a given DNF formula?
2. How many different variable assignments will satisfy a given 2-SAT formula?
3. How many perfect matchings are there for a given bipartite graph?

It was known quite a long time ago that finding a perfect matching for a given bipartite (undirected) graph $G = G(V, E)$ can be solved in polynomial $O(|V||E|)$ time, while the corresponding counting problem, "How many perfect matchings does the given bipartite graph have?" is already #P-complete. The problem of counting the number of perfect matchings is known to be equivalent to the computation of the permanent of a matrix [119].

Similarly, there is a trivial algorithm for determining whether a given DNF form of the satisfiability problem is true. Indeed, in this case we have to examine each clause, and if a clause is found that does not contain both a variable and its negation, then it is true, otherwise it is not. However, the counting version of this problem is #P-complete.

Many #P-complete problems have a *fully polynomial-time randomized approximation scheme* (FPRAS), which, informally, will produce with high probability an approximation to an arbitrary degree of accuracy, in time that is polynomial with respect to both the size of the problem and the degree of accuracy required [88]. Jerrum, Valiant, and Vazirani [60] showed that every #P-complete problem either has an FPRAS or is essentially impossible to approximate; if there is any polynomial-time algorithm that consistently produces an approximation of a #P-complete problem that is within a polynomial ratio in the size of the input of the exact answer, then that algorithm can be used to construct an FPRAS.

The use of fast decision-making algorithms (oracles) for solving NP-hard problems is very common in Monte Carlo methods; see, for example,

- Karger's paper [64], which presents an FPRAS for network unreliability based on the well-known DNF polynomial counting algorithm of Karp and Luby [65].
- The insightful monograph of Gertsbakh and Shpungin [41] in which Kruskal's spanning trees algorithm is used for estimating network reliability.

Our main strategy is as in [41]: *use fast polynomial decision-making oracles to solve #P-complete problems*. In particular, one can easily incorporate into SE the following:

- The breadth-first-search procedure or Dijkstra's shortest path algorithm [32] for counting the number of paths in the networks.
- The Hungarian decision-making assignment problem method [74] for counting the number of perfect matchings.
- The decision-making algorithm for counting the number of valid assignments in 2-SAT, developed in Davis et al. [35, 36].
- The Chinese postman (or Eulerian cycle) decision-making algorithm for counting the total number of shortest tours in a graph. Recall that in a Eulerian cycle problem, one must visit each edge at least once.

The remainder of this chapter is organized as follows. Section 5.2 presents some background on the classical OSLA algorithm and its extensions: (i) using an oracle and (ii) using multiple trajectories. Section 5.3 is the main discussion; it deals with both OSLA extensions simultaneously, resulting in the SE method.

Section 5.4 deals with applications of SE to counting in #P-complete problems, such as counting the number of trajectories in a network, number of satisfiability assignments in a SAT problem, and calculating the permanent. Here we also discuss how to choose the number of parallel trajectories in the SE algorithm. Unfortunately, there is no nSLA polynomial oracle for SAW. Consequently, SE is not directly applicable to SAW and we must resort to its simplified (time-consuming) version. This section describes the algorithms; some corresponding numerical results are given in Section 5.5. There we also show that SE outperforms the splitting method discussed in Chapter 4 and SampleSearch method of Gogate and Dechter [56, 57]. Our explanation for this is that SE is based on SIS (sequential importance sampling), whereas its two counterparts are based merely on the standard (nonsequential) importance sampling.

5.2 OSLA METHOD AND ITS EXTENSIONS

Let $|\mathcal{X}^*| < \infty$ be our counting quantity, such as,

- the number of satisfiability assignments in a SAT problem with n literals;
- the number of perfect matchings (permanent) in a bipartite graph with n nodes;
- the number of SAWs of length n.

We assume, similarly to in Chapter 4, that $\mathcal{X}^*$ is a subset of some larger but finite sample space $\mathcal{X}$. Let $f(\boldsymbol{x}) = 1/|\mathcal{X}|$ be the uniform pdf and $g(\boldsymbol{x})$ some arbitrary pdf on $\mathcal{X}$ that is positive for all $\boldsymbol{x} \in \mathcal{X}^*$, and suppose that we generate a random element $\boldsymbol{X} \in \mathcal{X}$ using these pdfs; then we can write

$$\begin{aligned}|\mathcal{X}^*| &= \frac{|\mathcal{X}^*|}{|\mathcal{X}|} \times |\mathcal{X}| = \mathbb{P}_f(\boldsymbol{X} \in \mathcal{X}^*) \times |\mathcal{X}| \\ &= \mathbb{E}_f[I_{\{\boldsymbol{X}\in\mathcal{X}^*\}}] \times |\mathcal{X}| = \mathbb{E}_g\left[I_{\{\boldsymbol{X}\in\mathcal{X}^*\}} \frac{f(\boldsymbol{X})}{g(\boldsymbol{X})}\right] \times |\mathcal{X}| \\ &= \mathbb{E}_g\left[I_{\{\boldsymbol{X}\in\mathcal{X}^*\}} \frac{1/|\mathcal{X}|}{g(\boldsymbol{X})}\right] \times |\mathcal{X}| = \mathbb{E}_g\left[\frac{I_{\{\boldsymbol{X}\in\mathcal{X}^*\}}}{g(\boldsymbol{X})}\right].\end{aligned}$$

This special case of importance sampling says that, when we generate a random element $\boldsymbol{X}$ using pdf $g(\boldsymbol{x})$, we define the (single-run) estimator

$$\widehat{|\mathcal{X}^*|} = \begin{cases} 1/g(\boldsymbol{X}), & \text{if } \boldsymbol{X} \in \mathcal{X}^*; \\ 0, & \text{if } \boldsymbol{X} \notin \mathcal{X}^*. \end{cases}$$

As usual, we take the sample mean of M independent retrials. Clearly, we obtain an unbiased estimator of the counting quantity, $\mathbb{E}_g[\widehat{|\mathcal{X}^*|}] = |\mathcal{X}^*|$. Furthermore, it is easy to see that the estimator has zero variance if and only if $g(\boldsymbol{x}) = I_{\{x \in \mathcal{X}^*\}}/|\mathcal{X}^*|$, the uniform pdf on the target set.

Next, we assume additionally that the sample space $\mathcal{X}$ is contained in some n-dimensional vector space; that is, samples $\boldsymbol{x} \in \mathcal{X}$ can be represented by vectors of n components. For instance, consider the problem of counting SAWs of length n. Each sample is a simple walk $\boldsymbol{x} = (x_1, \ldots, x_n)$ where the individual components x_t are points in the integer two-dimensional lattice. Then we might consider to decompose the importance sampling pdf $g(\boldsymbol{x})$ as a product of conditional pdfs:

$$g(\boldsymbol{x}) = g_1(x_1)\, g_2(x_2 \mid x_1) \cdots\, g_n(x_n \mid x_1, \ldots, x_{n-1}). \tag{5.1}$$

The importance sampling simulation of generating a random sample $\boldsymbol{X} \in \mathcal{X}$ is executed by generating sequentially the next component x_t $(t = 2, 3 \ldots, n)$ given the previous components $x_1, \ldots, x_{t-1}$. This method is called sequential importance sampling (SIS); see Section 5.7 in [108] for details in a general setting.

In particular, we apply a specific version of SIS, the OSLA procedure due to Rosenbluth and Rosenbluth [97], which was originally introduced for SAWs, as we described in the introductory section. Below, we give more details of the consecutive steps of the OSLA procedure, where we have in mind generating a sample $\boldsymbol{x} = (x_0, x_1, \ldots, x_n)$ for problems in which $\boldsymbol{x}$ represents a path of consecutive points $x_0, x_1, x_2, \ldots$. The first point x_0 of the path is fixed and constant for all paths, called the origin. For example, in the SAW problem, $x_0 = (0, 0)$, or in a network with source s and sink t, $x_0 = s$.

Procedure 1: ***OSLA***

1. **Initial step.** Start from the origin x_0. Set $t = 1$.
2. **Main step.** Let ν_t be the number of neighbors of x_{t-1} that have not yet been visited. If $x_t > 0$, choose x_t with probability $1/\nu_t$ from its neighbors. If $x_t = 0$, stop generating the path and deliver an estimate $\widehat{|\mathcal{X}^*|} = 0$.
3. **Stopping rule.** Stop if $t = n$. Otherwise, increase t by 1 and go to step 2.
4. **The estimator.** Return
$$\widehat{|\mathcal{X}^*|} = \nu_1 \ldots \nu_n. \tag{5.2}$$

Note that the OSLA procedure defines the SIS pdf:

$$g(\boldsymbol{x}) = \begin{cases} \dfrac{1}{\nu_1}\dfrac{1}{\nu_2}\cdots\dfrac{1}{\nu_n}, & \text{if } \boldsymbol{x} \in \mathcal{X}^*; \\ 0, & \text{if } \boldsymbol{x} \notin \mathcal{X}^*. \end{cases} \tag{5.3}$$

The OSLA counting algorithm now follows.

Algorithm 5.1 ***OSLA Algorithm***

1. *Generate independently M paths $\mathbf{X}_1, \ldots, \mathbf{X}_M$ from SIS pdf $g(\mathbf{x})$ via the above OSLA procedure.*
2. *For each path $\mathbf{X}_k$, compute the corresponding $\widehat{|\mathcal{X}^*|}_k$ as in (5.2). For the other parts (which do not reach the value n), set $\widehat{|\mathcal{X}^*|}_k = 0$.*
3. *Return the OSLA estimator:*

$$\widetilde{|\mathcal{X}^*|} = \frac{1}{M} \sum_{k=1}^{M} \widehat{|\mathcal{X}^*|}_k . \tag{5.4}$$

The efficiency of the OSLA method deteriorates rapidly as n becomes large. For example, it becomes impractical to simulate random SAWs of length more than 200. This is due to the fact that, if at any step t the point x_{t-1} does not have unoccupied neighbors ($\nu_t = 0$), then $\widehat{|\mathcal{X}^*|}$ is zero and it contributes nothing to the final estimate of $|\mathcal{X}^*|$.

As an example, consider again Figure 5.2, depicting a SAW trapped after 15 iterations. One can easily see that the corresponding values ν_t, $t = 1, \ldots, 15$ of each of the 15 points are

$$\begin{array}{llllllll} \nu_1 = 4, & \nu_2 = 3, & \nu_3 = 3, & \nu_4 = 3, & \nu_5 = 3, & \nu_6 = 3, & \nu_7 = 2, & \nu_8 = 3, \\ \nu_9 = 3, & \nu_{10} = 3, & \nu_{11} = 2, & \nu_{12} = 3, & \nu_{13} = 2, & \nu_{14} = 1, & \nu_{15} = 0. \end{array}$$

As for another situation where OSLA can be readily trapped, consider a directed graph in Figure 5.3 with source s and sink t. There are two ways from s to t, one via nodes $a_1, \ldots, a_n$ and another via nodes $b_1, \ldots, b_n$. Figure 5.3 corresponds to $n = 3$. Note that all nodes besides s and t are directly connected to a central node o, which has no connection with t. Clearly, in this case, most of the random walks (with probability $1 - (1/2)^n$) will be trapped at node o.

5.2.1 Extension of OSLA: *n*SLA Method

A natural extension of the OSLA procedure is to look ahead more than one step. We consider only the case of looking all the way to the end, which we call the

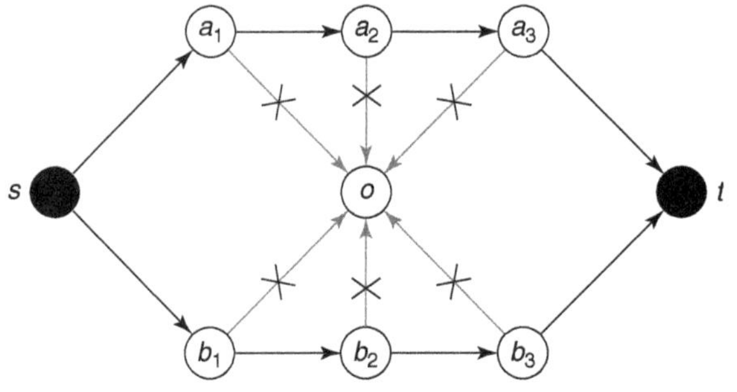

Figure 5.3 Directed graph.

nSLA method. This means that no path will be ever lost, and it implies that $\nu_t > 0$, for all t.

Consider, for example, the SAW in Figure 5.2. If we could use the nSLA method instead of OSLA, we would rather move after step 13 (corresponding to point x_{13}) South instead of North. By doing so we would prevent the SAW being trapped after 15 iterations.

For many problems, including SAWs, the nSLA algorithm requires additional memory and computing power; for that reason, it has limited applications. There exists, however, a set of problems where nSLA can be implemented easily by using polynomial-time oracles. As mentioned, relevant examples are counting perfect matchings in a graph and counting the number of paths in a network.

The nSLA procedure does not differ much from the OSLA procedure. For completeness, we present it below.

Procedure 2: ***n*SLA**

1. **Initial step.** Same as in **Procedure 1**.
2. **Main step.** Employ an oracle and find $\nu_t^* \geq 1$ the number of neighbors of x_{t-1} that have not yet been visited and from which the path can be completed successfully. Choose x_t with probability $1/\nu_t^*$ from its neighbors.
3. **Stopping rule.** Same as in **Procedure 1**.
4. **The Estimator.** Same as in **Procedure 1**.

To see how nSLA works in practice, consider a simple example following the main steps of the OSLA Algorithm 5.1.

EXAMPLE 5.1

Let $\mathcal{X} = \{0, 1\}^3$ be the sample space consisting of the three-dimensional binary vectors, and suppose that $\mathcal{X}^* = \{000, 001, 100, 110, 111\}$ is the target set of valid combinations (or paths). We have $n = 3$ and $|\mathcal{X}^*| = 5$. Note that in the nSLA algorithm, it always holds that $\nu_t^* \geq 1$, $t = 1, 2, 3$, whereas $\nu_t = 0$ may occur in the OSLA algorithm. Figure 5.4 presents a tree corresponding to the set $\{000, 001, 100, 110, 111\}$. In this toy example, there is no need to consider the origin x_0.

According to nSLA **Procedure 2**, we start with the first binary variable x_1. Because $x_1 \in \{0, 1\}$, we employ the oracle two times; once for $x_1 = 0$ and once for $x_1 = 1$. That is, we ask the oracle whether there is a valid combination among all triplets $0x_2x_3$; also we ask this question concerning the triplets $1x_2x_3$. The role of the oracle is merely to provide a YES-NO answer for these questions. Clearly, in our case, the answer is YES in both cases, because an example of $0x_2x_3$ is, say, 000 and an example of $1x_2x_3$ is, say, 110. We therefore have $\nu_1^* = 2$. Following

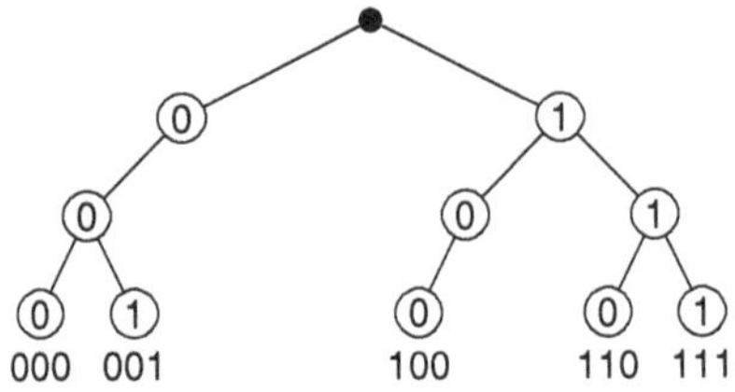

Figure 5.4 Tree corresponding to the set {000, 001, 100, 110, 111}.

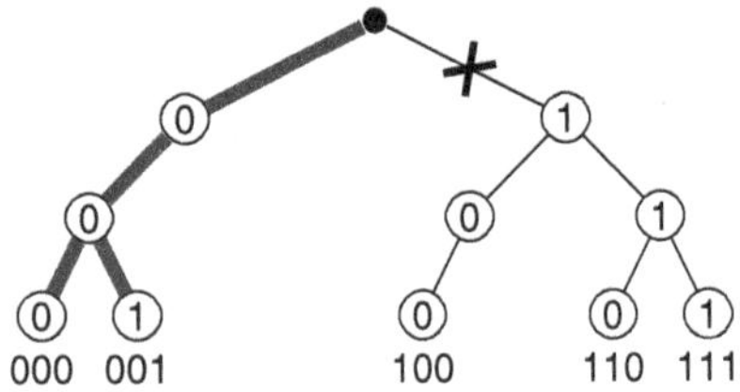

Figure 5.5 The subtrees {000, 001} (in bold) generated by nSLA using the oracle.

Procedure 2, we next flip a symmetric coin. Assume that the outcome is 0. This means that we will proceed to path $0x_2x_3$.

Consider next $x_2 \in \{0, 1\}$. Again, we employ the oracle two times; once for $x_2 = 0$ (by considering the triplet $00x_3$) and once for $x_2 = 1$ (by considering the triplet $01x_3$). In this case, the answer is YES for the case $00x_3$ (an example of $00x_3$ is, as before, 000) and NO for the case $01x_3$ (there is no valid assignment in the set {000, 001, 100, 110, 111} for $01x_3$ with $x_3 \in \{0, 1\}$). We therefore have $\nu_2^* = 1$, and we proceed to path $00x_3$.

We finally ask the oracle about x_3: do $x_3 = 0$ and $x_3 = 1$ yield feasible paths? In this case, the answer is YES for both cases, because we automatically obtain 000 and 001. We have $\nu_3^* = 2$. The resulting estimator is therefore

$$\widehat{|\mathcal{X}^*|} = \nu_1^* \nu_2^* \nu_3^* = 2 \cdot 1 \cdot 2 = 4.$$

Figure 5.5 presents the subtrees {000, 001} (in bold) generated by nSLA using the oracle.

Similarly to the analysis above for the trajectories 000 and 001, the valid trajectory 100 results in $\widehat{|\mathcal{X}^*|} = 4$, whereas the valid trajectories 110, 111 result in $\widehat{|\mathcal{X}^*|} = 8$. Because the corresponding probabilities for $\widehat{|\mathcal{X}^*|} = 4$ and $\widehat{|\mathcal{X}^*|} = 8$ are $1/4$ and $1/8$, we can compute the variance of $\widehat{|\mathcal{X}^*|}$:

$$\mathbb{V}\mathrm{ar}[\widehat{|\mathcal{X}^*|}] = 1/5\{3(4-5)^2 + 2(8-5)^2\} = 21/2.$$

As a comparison, the OSLA algorithm generates the valid combinations each with probability $g(\boldsymbol{x}) = 1/8$ (with probability $3/8$ OSLA is trapped). Hence the associated OSLA estimator has variance 15. □

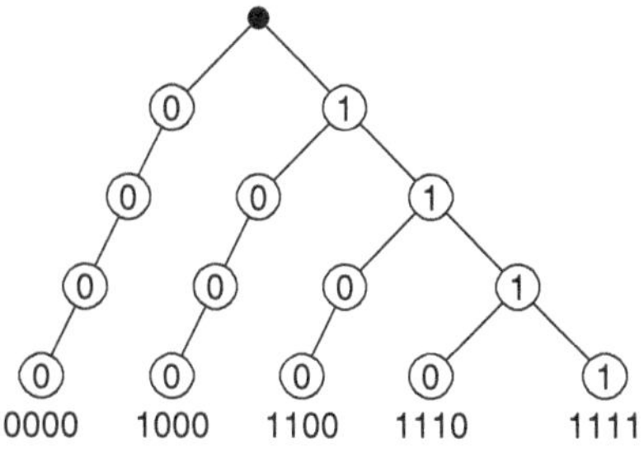

Figure 5.6 A graph with four literals and $|\mathcal{X}^*| = 5$.

The main drawback of nSLA **Procedure 2** is that its SIS pdf $g(\boldsymbol{x})$ is model dependent. Typically, it is far from the ideal uniform SIS pdf $g^*(\boldsymbol{x})$, that is,

$$g(\boldsymbol{x}) \neq g^*(\boldsymbol{x}) = \frac{1}{|\mathcal{X}^*|}.$$

As a result, the estimator of $|\mathcal{X}^*|$ has a large variance.

To see the nonuniformity of $g(\boldsymbol{x})$, consider a 2-SAT model with clauses $C_1 \wedge C_2 \wedge \cdots \wedge C_{n-1}$, where $C_i = x_i \wedge \quad \bar{x}_{i+1} \geq 1, \quad i = 1, \ldots, n-1$. Figure 5.6 presents the corresponding graph with four literals and $|\mathcal{X}^*| = 5$. In this case, $|\mathcal{X}^*| = n + 1$, the zero variance pdf $g^*(\boldsymbol{x}) = 1/|\mathcal{X}^*| = 1/(n+1)$, whereas the SIS pdf is

$$g(\boldsymbol{x}) = \begin{cases} 1/2, & \text{for } (00, \ldots, 00), \\ 1/2^2, & \text{for } (10, \ldots, 00), \\ 1/2^3, & \text{for } (11, \ldots, 00), \\ 1/2^n, & \text{for } (11, \ldots, 10) \text{ and } (11, \ldots, 11), \end{cases}$$

which is highly nonuniform.

To improve the nonuniformity of $g(\boldsymbol{x})$, we will run the nSLA procedure in parallel by multiple trajectories. Before doing so, we present below for convenience a multiple trajectory version of the OSLA Algorithm 5.1 for SAWs.

5.2.2 Extension of OSLA for SAW: Multiple Trajectories

In this section, we extend the OSLA method to multiple trajectories and present the corresponding algorithm. For ease of referencing, we call it multiple OSLA. We provide it here for counting SAWs.

5.2.2.1 Multiple-OSLA Algorithm for SAW

Recall that a SAW is a sequence of moves in a lattice such that it does not visit the same point more than once. There are specially designed counting algorithms, for instance, the pivot algorithm [30]. Empirically we found that these outperform our multiple-OSLA algorithm; nevertheless, we present it below in the interest of clarity of representation and for motivational purposes.

For simplicity, we assume that the walk starts at the origin, and we confine ourselves to the two-dimensional integer lattice case. Each SAW is represented by a path $\boldsymbol{x} = (x_1, x_2, \ldots, x_{n-1}, x_n)$, where $x_i \in \mathbb{Z}^2$. A SAW of length $n = 130$ has been given in Figure 5.1.

Algorithm 5.2 ***Multiple OSLA for SAW***

1. ***Iteration 1***
   - ***Full Enumeration.*** *Select a small number n_0, say $n_0 = 4$ and count via full enumeration all different SAWs of size n_0 starting from the origin $(0, 0)$. Denote the total number of these SAWs by $N_1^{(e)}$ and call them the elite sample. For example, for $n_0 = 4$, the number of elites $N_1^{(e)} = 100$. Set the first level to n_0. Proceed with the $N_1^{(e)}$ elites from level n_0 to the next one, $n_1 = n_0 + r$, where r is a small integer (typically $r = 1$ or $r = 2$) and count via full enumeration all different SAWs at level $n_1 = n_0 + r$. Denote the total number of such SAWs by N_1. For example, for $n_1 = 5$ there are $N_1 = 284$ different SAWs.*
   - ***Calculation of the First Weight Factor.*** *Compute*

$$\nu_1 = \frac{N_1}{N_1^{(e)}}, \tag{5.5}$$

   and call it the first weight factor.

2. ***Iteration*** t, $t \geq 2$
   - ***Full Enumeration.*** *Proceed with $N_{t-1}^{(e)}$ elites from iteration $t - 1$ to the next level $n_{t-1} = n_{t-2} + r$ and derive via full enumeration all SAWs at level $n_{t-1} = n_{t-2} + r$, that is, of all those SAWs that continue the $N_{t-1}^{(e)}$ paths resulting from the previous iteration. Denote by N_t the resulting number of such SAWs.*
   - ***Stochastic Enumeration.*** *Select randomly without replacement $N_t^{(e)}$ SAWs from the set of N_t ones and call this step stochastic enumeration.*
   - ***Calculation of the t-th Weight Factor.*** *Compute*

$$\nu_t = \frac{N_t}{N_t^{(e)}}, \tag{5.6}$$

   and call it the t-th weight factor. Similar to OSLA, the weight factor ν_t can be both ≥ 1 and $= 0$; in case$\nu_t = 0$, we stop and deliver $\widehat{|\mathcal{X}^*|} = 0$.

3. ***Stopping Rule***
   - *Proceed with iteration t, $t = 1, \ldots, \frac{n-n_0}{r}$ and compute*

$$\widehat{|\mathcal{X}^*|} = N_1^{(e)} \prod_{t=1}^{(n-n_0)/r} \nu_t. \tag{5.7}$$

   - *Call $\widehat{|\mathcal{X}^*|}$ the point estimator of $|\mathcal{X}^*|$.*

- *Since the number of levels is fixed,* $\widehat{|\mathcal{X}^*|}$ *presents an unbiased estimator of* $|\mathcal{X}^*|$ *(see [73]).*

4. ***Final Estimator***
   - *Run Steps 1–3 for M independent replications and deliver*

$$\widetilde{|\mathcal{X}^*|} = \frac{1}{M}\sum_{k=1}^{M} \widehat{|\mathcal{X}^*|}_k \tag{5.8}$$

   as an unbiased estimator of $|\mathcal{X}^*|$.
   - *Call* $\widetilde{|\mathcal{X}^*|}$ *the multiple-OSLA estimator of* $|\mathcal{X}^*|$.

A few remarks can be made:

- Typically, one keeps the number of elites $N_t^{(e)}$ in multiple OSLA fixed, say $N_t^{(e)} = N^{(e)} = 100$, while N_t varies from iteration to iteration.
- OSLA represents a particular case of multiple OSLA, when the number of elites is $N_t^{(e)} = 1$ in all iterations.
- In contrast to OSLA, which loses most of its trajectories (even for modest n), with multiple OSLA we can reach quite large levels, say $n = 10{,}000$, provided the number $N_t^{(e)}$ of elite samples is not too small, say $N_t^{(e)} = 100$ in all iterations. Hence, one can generate very long random SAWs with multiple OSLA.
- The sample variance of $\widehat{|\mathcal{X}^*|}$ is

$$S^2(\widetilde{|\mathcal{X}^*|}) = \frac{1}{M-1}\sum_{k=1}^{M} (\widehat{|\mathcal{X}^*|}_k - \widetilde{|\mathcal{X}^*|})^2, \tag{5.9}$$

  and the relative error is

$$\mathrm{RE}(\widetilde{|\mathcal{X}^*|}) = \frac{S(\widetilde{|\mathcal{X}^*|})}{\widetilde{|\mathcal{X}^*|}}. \tag{5.10}$$

- There is similarity with the splitting method of Chapter 4 in the sense that the whole state space is broken in smaller search spaces, in which the promising samples (elites) are chosen for prolongation.

EXAMPLE 5.2

The first two iterations of Algorithm 5.2 for $n_0 = 1$, $r = 1$, and $N_t^{(e)} = N^{(e)} = 4$ are given below.

1. **Iteration 1**
   - **Full Enumeration.** We set $n_0 = 1$ and count (via full enumeration) all different SAWs of length n_0 starting from the origin $(0, 0)$. We have $N_1^{(e)} = 4$ (see Figure 5.7). We proceed (again via full enumeration) to derive from the $N_1^{(e)} = 4$ elites all SAWs of length $n_1 = 2$ (there are $N_1 = 12$ of them, see part (a) of Figure 5.8).

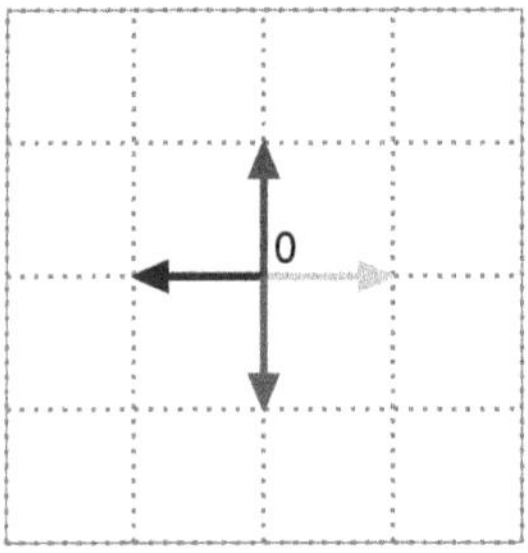

Figure 5.7 The first four elites $N_1^{(e)} = N^{(e)} = 4$.

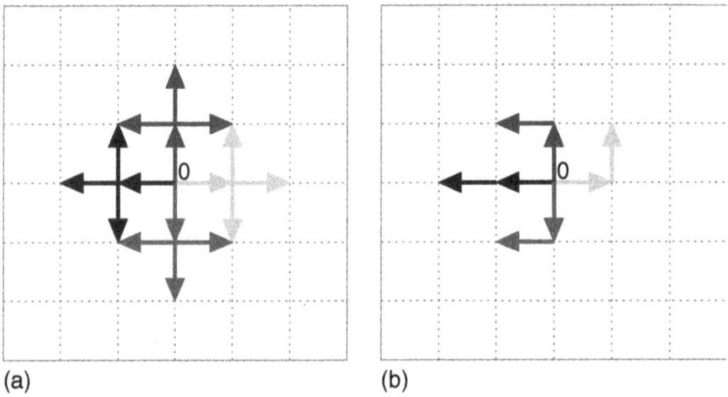

Figure 5.8 First iteration of Algorithm 5.2.

- **Calculation of the first weight factor.** We have $\nu_1 = N_1/N_1^{(e)} = 12/4 = 3$.

2. **Iteration 2**
   - **Stochastic Enumeration.** Select randomly without replacement $N_2^{(e)} = 4$ elites from the set of $N_1 = 12$ ones (see part (b) of Figure 5.8).
   - **Full Enumeration.** Proceed via full enumeration to determine all SAWs of length $n_2 = 3$ that extend the $N_2^{(e)} = 4$ elites. There are $N_2 = 12$ of them, see part (a) of Figure 5.9.
   - **Calculation of the second weight factor.** We have $\nu_2 = N_2/N_2^{(e)} = 12/4 = 3$.

Remark 5.1

Here we support empirically a previous remark that multiple OSLA can generate very long random SAWs. To do so, we ran the multiple OSLA algorithm for several fixed values of $N_t^{(e)} = N^{(e)}$ and $r = 1$. Specifically, we considered the values $N^{(e)} = 1, 2, 5, 15, 25, 35$. For small elite sizes, the full enumeration numbers $N_t, t = 1, 2, \ldots$ are also small (relatively), and easy to compute. For instance a run with $N^{(e)} = 2$ resulted in a trajectory of

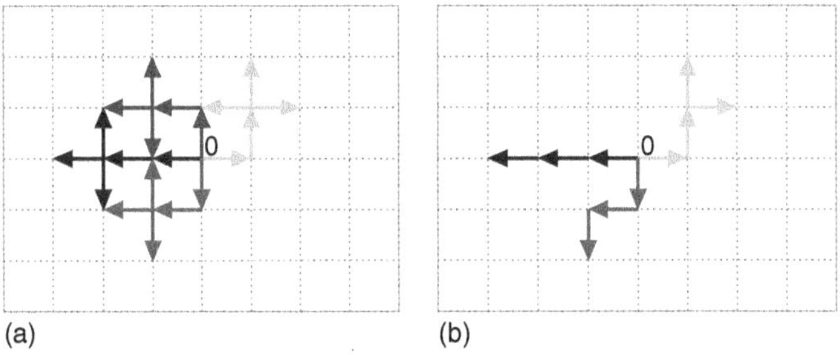

Figure 5.9 Second iteration of Algorithm 5.2.

Table 5.1 Ordered lengths $R_{k,N^{(e)}}$ of simulated SAW's for $N^{(e)}$=1, 2, 5, 15, 25, 35

k	$R_{k,1}$	$R_{k,2}$	$R_{k,5}$	$R_{k,15}$	$R_{k,25}$	$R_{k,35}$
1	10	26	73	322	791	1185
5	33	62	193	539	1757	2057
10	44	178	306	1205	3185	2519
15	79	299	436	1849	4099	5537
20	198	644	804	4760	6495	9531
Average	64.2	216.25	356.6	1503.25	3343	4080.6

length 30 with consecutive N_t-numbers

$$6, 6, 6, 6, 6, 6, 5, 5, 5, 4, 4, 6, 5, 6, 6$$
$$6, 5, 4, 6, 6, 6, 5, 4, 4, 3, 4, 3, 2, 2, 0$$

For the $N^{(e)}$ values mentioned above, we executed the algorithm 20 times independently, each time until the algorithm got stuck. The observed 20 lengths of SAW trajectories were ordered and denoted by $R_{k,N^{(e)}}$ ($k = 1$ for the smallest, $k = 20$ for the largest length). Table 5.1 shows 5 of these 20 lengths (per $N^{(e)}$) and the average of all 20 lengths.

The first R-column of Table 5.1 reflects simulation results for OSLA. The average length of the generated SAW trajectories is about 65. After many more experiments with OSLA, we found that only 1% of the generated trajectories reaches length 200.

We denote by $\overline{R}_m$ the average length of the simulated trajectories using elite size $m = N^{(e)}$, as shown in the last row of the table for $m = 1, 2, 5, 15, 25, 35$. It is interesting to plot the regression line of $\{(m, \overline{R}_m) : m = 1, 2, \ldots\}$ based on the data $m = 1, 2, 5, 15, 25, 35$. We see that the average trajectory length increases linearly with elite size (Figure 5.10).

Experiments with the multiple-OSLA algorithm for general counting problems, like SAT, indicated that this method is also susceptible to generating many zeros (trajectories of zero length); even for a relatively small model sizes. For this reason it has limited applications.

For example, consider a 3-SAT model with an adjacency matrix $\boldsymbol{A} = (20 \times 80)$ and $|\mathcal{X}^*| = 15$. Running the multiple-OSLA Algorithm 5.2 with $N^{(e)} = 2$, we

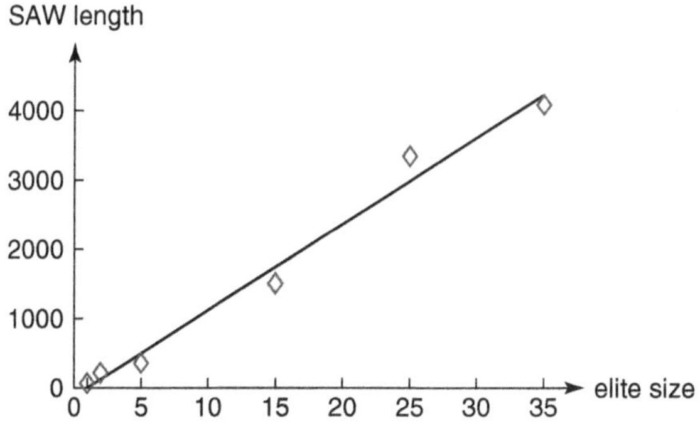

Figure 5.10 Regression of SAW trajectory lengths against elite size.

observed that it discovered all 15 valid assignments (satisfying all 80 clauses) only with probability $1.4 \cdot 10^{-4}$. We found that, as the size of the models increases, the percentage of valid assignments goes rapidly to zero.

5.3 SE METHOD

Here we present our main algorithm, called *stochastic enumeration* (SE), which extends both the nSLA and the multiple-OSLA algorithms as follows:

- SE extends nSLA in the sense that it uses multiple trajectories instead of a single one.
- SE extends the multiple-OSLA Algorithm 5.2 in the sense that it uses an oracle at the **Full Enumeration** step.

We present SE for the models where the components of the vector $\boldsymbol{x} = (x_1, \ldots, x_n)$ are assumed to be binary variables, such as SAT, and where fast decision-making nSLA oracles are available (see [35, 36] for SAT). Its modification to arbitrary discrete variables is straightforward.

5.3.1 SE Algorithm

Consider the multiple-OSLA Algorithm 5.2 and assume for simplicity that $r = 1$ and that at iteration $t - 1$ the number of elites is, say, $N_{t-1}^{(e)} = 100$. In this case, it can be seen that, in order to implement an oracle in the Full Enumeration step at iteration t, we have to run it $2N_{t-1}^{(e)} = 200$ times: 100 times for $x_t = 0$ and 100 more times for $x_t = 1$. For each fixed combination of $(x_1, \ldots, x_t)$, the oracle can be viewed as an nSLA algorithm in the following sense:

- It sets first $x_{t+1} = 0$ and then makes a decision (YES-NO path) for the remaining $n - t + 1$ variables $(x_{t+2}, \ldots, x_n)$.
- It sets next $x_{t+1} = 1$ and then again makes a similar (YES-NO) decision for the same set $(x_{t+2}, \ldots, x_n)$.

Algorithm 5.3 *SE Algorithm*

1. ***Iteration 1***
   - ***Full Enumeration.*** *(Similar to Algorithm 5.2.) Let* n_0 *be the number corresponding to the first* n_0 *variables* $x_1, \ldots, x_{n0}$. *Count via full enumeration all different paths (valid assignments in SAT) of size* n_0. *Denote the total number of these paths (assignments) by* $N_1^{(e)}$ *and call them the elite sample. Proceed with the* $N_1^{(e)}$ *elites from level* n_0 *to the next one* $n_1 = n_0 + r$, *where* r *is a small integer (typically* $r = 1$ *or* $r = 2$*) and count via full enumeration all different paths (assignments) of size* $n_1 = n_0 + r$. *Denote the total number of such paths (assignments) by* N_1.

   - ***Calculation of the First Weight Factor.*** *Same as in Algorithm 5.2.*
2. ***Iteration*** $t, t \geq 2$
   - ***Full Enumeration.*** *(Same as in Algorithm 5.2, except that it is performed via the corresponding polynomial time oracle rather than OSLA.) Recall that for each fixed combination of* $(x_1, \ldots, x_t)$, *the oracle can be viewed as an* $(n - t + 1)$*-step look-ahead algorithm in the sense that it*
     - *Sets first* $x_{t+1} = 0$ *and then makes a YES-NO decision for the path associated with the remaining* $n - t + 1$ *variables* $(x_{t+2}, \ldots, x_n)$.
     - *Sets next* $x_{t+1} = 1$ *and then again makes a similar (YES-NO) decision.*
   - ***Stochastic Enumeration.*** *Same as in Algorithm 5.2.*
   - ***Calculation of the*** t***-th Weight Factor.*** *(Same as in Algorithm 5.2.) Recall that, in contrast to multiple OSLA, where* $v_t \geq 0$, *here* $v_t \geq 1$.
3. ***Stopping Rule.*** *Same as in Algorithm 5.2.*
4. ***Final Estimator.*** *Same as in Algorithm 5.2.*

It follows that, if $N_t^{(e)} \geq |\mathcal{X}^*|$ in all iterations, the SE Algorithm 5.3 will be exact.

Remark 5.2

The SE method can be viewed as a multilevel splitting method with a specially chosen importance function [48]. In the traditional splitting method [12] and [13, 104], one chooses an obvious importance function, say the number of satisfied clauses in a SAT problem or the length of a walk in SAW. Here we simply decompose the rare event into non rare events by introducing intermediate levels. In the proposed method, we introduce a random walk on a graph, starting at some node on the graph, and we try to find an alternative (better) importance function that can be computed in polynomial time and that provides a reasonable estimate of the probability of the rare event. This is the same as saying that we are choosing a specific dynamics on the graph, and we are trying to optimize the importance function for this precise dynamics.

It is well known that the best possible importance function in a dynamical rare-event environment that is driven by a nonstationary Markov chain is the committor function [87]. It is also well known that its computation is at least as difficult as the underlying rare

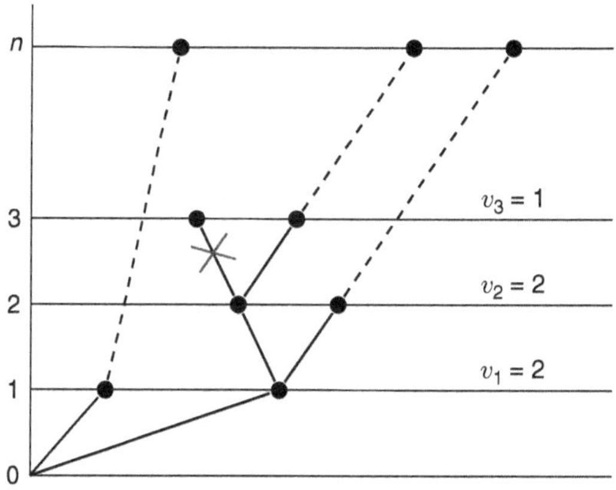

Figure 5.11 Dynamics of the SE Algorithm 5.3 for the first three iterations.

event. In the SE approach, however, we roughly approximate the committor function using an oracle in the sense that we can say whether this probability is zero or not. For example, in the SAT problem with already assigned literals, we can compute whether or not they can lead to a valid solution. The committor is equal to the probability $z \in [0, 1]$ of hitting the rare event as a function of the initial state x_0 of the Markov chain. In particular, for SAT, the committor being 0 implies that we keep 0, otherwise we take 1; for SAW, we may also have 1 on points for which the committor is 0.

Figure 5.11 presents the dynamics of the SE Algorithm 5.3 for the first three iterations in a model with n variables using $N^{(e)} = 1$. There are three valid paths (corresponding to the dashed lines) reaching the final level n (with the aid of an oracle), and one invalid path (indicated by the cross). It follows from Figure 5.11 that the accumulated weights are $v_1 = 2, v_2 = 2, v_3 = 1$.

The SE estimator $\widehat{|\mathcal{X}^*|}$ is unbiased for the same reason as its multiple-OSLA counterpart (5.7); namely, both can be viewed as multiple splitting methods with fixed (nonadaptive) levels [16].

Below we show how SE works for several toy examples.

EXAMPLE 5.3

Consider the SAT model

$$(x_1 \vee x_2) \wedge (x_2 \vee x_3) \wedge (x_3 \vee x_4).$$

Suppose that $n_0 = 2$, $r = 1$, $N_t^{(e)} = 3$, and $M = 1$.

1. **Iteration 1**
   - **Full Enumeration.** Since $n_0 = 2$, we handle first the variables x_1 and x_2. Using SAT solver we obtain three trajectories ($01x_3x_4$, $10x_3x_4$, $11x_3x_4$), which can be extended to valid solutions. Trajectory $00x_3x_4$ cannot be

extended to a valid solution and is discarded by the oracle. We have, therefore, $N_1^{(e)} = 3$, which is still within the allowed budget $N_t^{(e)} = 3$.

We proceed to x_3 and, by using the oracle, we derive from the $N_1^{(e)} = 3$ elites all SAT assignments of length $n_1 = 3$. By full enumeration we obtain the trajectories $(011x_4,\ 010x_4,\ 101x_4,\ 110x_4,\ 111x_4)$. Trajectory $100x_4$ cannot be extended to a valid solution and is discarded by the oracle. We have, therefore, $N_1 = 5$.

- **Calculation of the first weight factor.** We have $\nu_1 = N_1/N_1^{(e)} = 5/3$.

2. **Iteration 2**
   - **Stochastic Enumeration.** Because $N_1 > N_t^{(e)} = 3$ we resort to sampling by selecting randomly without replacement $N_2^{(e)} = 3$ trajectories from the set of $N_1 = 5$. Suppose we pick $(010x_4,\ 101x_4,\ 111x_4)$. These will be our working trajectories in the next step.
   - **Full Enumeration.** We proceed with the oracle to handle x_4; that is, derive from the $N_1^{(e)} = 3$ elites all valid SAT assignments of length $n_2 = 4$. By full enumeration we obtain the trajectories (0101, 1010, 1011, 1110, 1111). We have, therefore, again $N_2 = 5$.
   - **Calculation of the second weight factor.** We have $\nu_2 = N_2/N_2^{(e)} = 5/3$.

The SE estimator of the true $|\mathcal{X}^*| = 8$ based on the above two iterations is $\widehat{|\mathcal{X}^*|} = 3 \cdot 5/3 \cdot 5/3 = 25/3$. When we would take a larger number of elites in each iteration, say $N_t^{(e)} = 5$, we would get the exact result: $\widehat{|\mathcal{X}^*|} = 3 \cdot 5/3 \cdot 8/5 = 8$. □

Example 5.3 is the $n = 4$ version of a special type of 2-SAT problems involving n literals and $n - 1$ clauses:

$$C_1 \wedge C_2 \wedge \cdots \wedge C_{n-1}, \quad \text{where } C_i = x_i \wedge x_{i+1}, \quad i = 1, \ldots, n-1.$$

For any initial number n_0 of literals in the first iteration, we have $N_1^{(e)} = F_{n_0-2}$ and $N_1 = F_{n_0-1}$, where F_n denotes the n-th Fibonacci number. In particular, if $n_0 = 12$, we have $N_1^{(e)} = F_{10} = 88$ and $N_1 = F_{11} = 133$ different SAT assignments.

EXAMPLE 5.4

Consider the SAT model

$$(x_1 \vee x_2) \wedge (x_2 \vee x_3) \wedge (x_2 \vee x_4).$$

Suppose again that $n_0 = 2$, $r = 1$, $N_t^{(e)} = 3$, and $M = 1$.

1. **Iteration 1** The same as in Example 5.3.
2. **Iteration 2**
   - **Stochastic Enumeration.** We select randomly without replacement $N_2^{(e)} = 3$ elites from the set of $N_1 = 5$. Suppose we pick $(010x_4,\ 110x_4,\ 111x_4)$.

- **Full Enumeration.** We proceed with oracle to handle x_4, that is, derive from the $N_1^{(e)} = 3$ elites all SAT assignments of length $n_2 = 4$. By full enumeration we obtain the trajectories $(0100, 0101, 1100, 1101, 1110, 1111)$. We have, therefore, $N_2 = 6$.
- **Calculation of the second weight factor.** We have $\nu_2 = N_2/N_2^{(e)} = 2$.

The estimator of the true $|\mathcal{X}^*| = 9$ based on the above two iterations is $\widehat{|\mathcal{X}^*|} = 3 \cdot 5/3 \cdot 2 = 10$. It is readily seen that if we set again $N_t^{(e)} = 5$ instead of $N_t^{(e)} = 3$, we would get the exact result, that is, $|\mathcal{X}^*| = 3 \cdot 5/3 \cdot 9/5 = 9$. □

One can see that as $N^{(e)}$ increases, the variance of the SE estimator $\widehat{|\mathcal{X}^*|}$ in (5.7) decreases and for $N^{(e)} \geq \widehat{|\mathcal{X}^*|}$ we have $\mathbb{V}\mathrm{ar}[\widehat{|\mathcal{X}^*|}] = 0$.

EXAMPLE 5.5 ***Example 5.1 Continued***

Let again $\boldsymbol{x} = (x_1, x_2, x_3)$ be a three dimensional vector with $\{000, 001, 100, 110, 111\}$ being the set of its valid combinations. As before, we have $n = 3$, $|\mathcal{X}_0| = 2^3 = 8$ and $|\mathcal{X}^*| = 5$. In contrast to Example 5.1, where $N^{(e)} = 1$, we assume that $N^{(e)} = 2$.

We have $N_1 = 3$ and $\nu_1 = 3/2$. Because $N_k^{(e)} = N^{(e)} = 2$, $k = 1, 2, 3$ we proceed with the following three pairs $(00, 10)$, $(00, 11)$, $(10, 11)$. They result into $(000, 001, 100)$, $(000, 001, 110, 111)$, $(100, 110, 111)$, respectively, with their corresponding pairs $(N_2 = 3, \nu_2 = 3/2), (N_2 = 4, \nu_2 = 2), (N_2 = 3, \nu_2 = 3/2)$.

It is readily seen that the estimator of $|\mathcal{X}^*|$ based on $(000, 001, 100)$ and $(100, 110, 111)$ equals $\widehat{|\mathcal{X}^*|} = 2 \cdot 3/2 \cdot 3/2 = 9/2$, and the one based on $(000, 001, 110, 111)$ is $\widehat{|\mathcal{X}^*|} = 6$. Noting that their probabilities are equal to 1/3 and averaging over all three cases we obtain the desired results $|\mathcal{X}^*| = 5$. The variance of $\widehat{|\mathcal{X}^*|}$ is

$$\mathbb{V}\mathrm{ar}[\widehat{|\mathcal{X}^*|}] = 1/3\{2(9/2 - 5)^2 + 2(6 - 5)^2\} = 1/2.$$

It follows from the above that by increasing $N^{(e)}$ from 1 to 2, the variance decreases 21 times. Figure 5.12 presents the subtrees $\{000, 001, 100\}$ (in bold) of the original tree (based on the set $\{000, 001, 100, 110, 111\}$), generated using the oracle with $N^{(e)} = 2$. □

EXAMPLE 5.6

Consider a problem with five binary variables and solution set represented by Figure 5.13. Figures 5.14 and 5.15 present typical subtrees (in bold) generated by setting $N^{(e)} = 1$, and $N^{(e)} = 2$, respectively. □

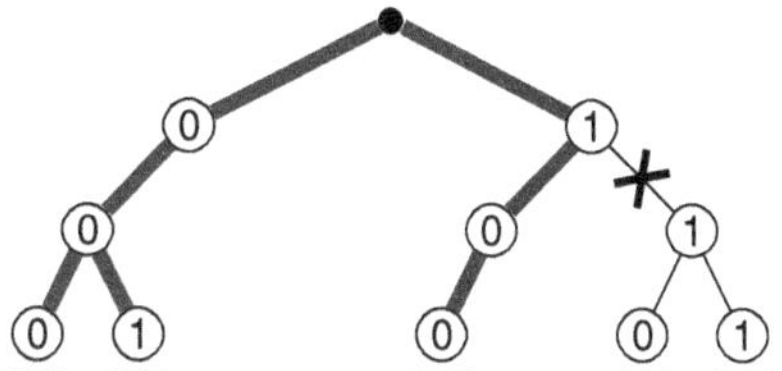

Figure 5.12 The subtrees {000, 001, 100} (in bold) corresponding to $N^{(e)} = 2$ in Example 5.5.

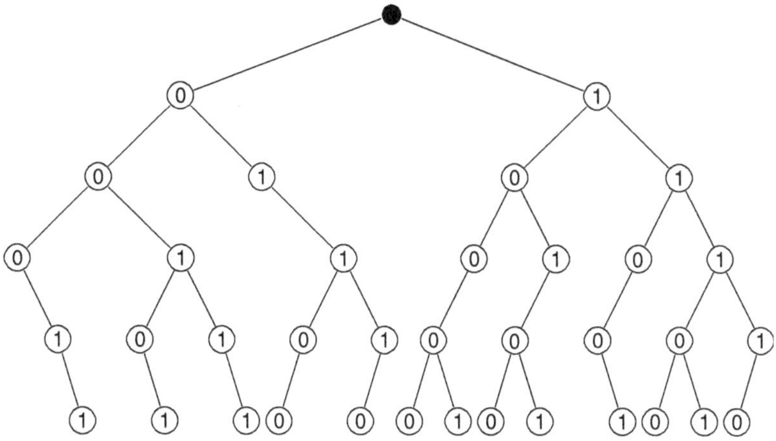

Figure 5.13 The problem tree of Example 5.5 having five variables.

As mentioned earlier, the major advantage of SE Algorithm 5.3 versus its nSLA counterpart is that the uniformity of $g(\boldsymbol{x})$ in the former increases in $N^{(e)}$. In other words, $g(\boldsymbol{x})$ becomes "closer" to the ideal pdf $g^*(\boldsymbol{x})$. We next demonstrate numerically this phenomenon while considering the following two simple models:

i. A 2-SAT model of n literals with clauses $C_1 \wedge C_2 \wedge \cdots \wedge C_{n-1}$, where $C_i = x_i \vee \bar{x}_{i+1}, \quad i = 1, \ldots, n-1$. It is easy to check that $|\mathcal{X}^*| = n + 1$. See Figure 5.6 for $n = 4$ literals and $|\mathcal{X}^*| = 5$. Straightforward calculation yields that, for this particular case, the variance reduction obtained by using $N^{(e)} = 2$ instead of $N^{(e)} = 1$, is about 150 times.
Table 5.2 presents the efficiency of the SE Algorithm 5.3 for the model having $|\mathcal{X}^*| = 100$. We considered different values of $N^{(e)}$ and M. The comparison was done for

$$(N^{(e)} = 1,\ M = 500),\ (N^{(e)} = 5,\ M = 100),\ (N^{(e)} = 10,\ M = 50),$$
$$(N^{(e)} = 25,\ M = 20),\ (N^{(e)} = 50,\ M = 10)\ (N^{(e)} = 75,\ M = 5),$$
$$(N^{(e)} = 100,\ M = 1).$$

The relative error RE was calculated as

$$\mathrm{RE} = \frac{\left(1/10 \sum_{i=1}^{10} (|\widetilde{\mathcal{X}^*}|_i - 100)^2\right)^{1/2}}{100}.$$

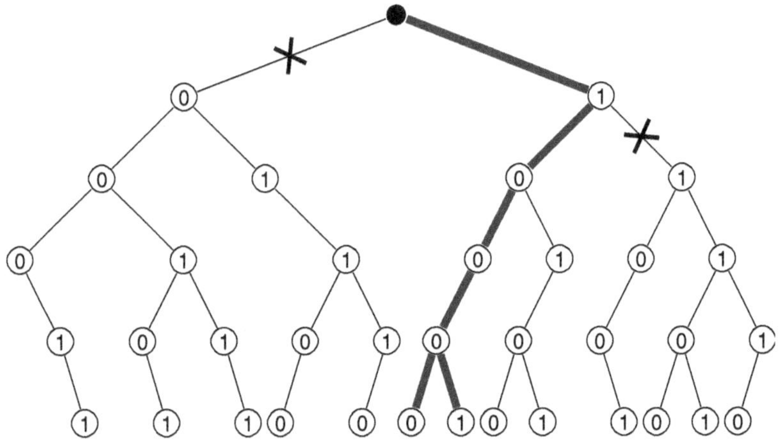

Figure 5.14 Subtree (in bold) corresponding to $N^{(e)} = 1$ in Example 5.6.

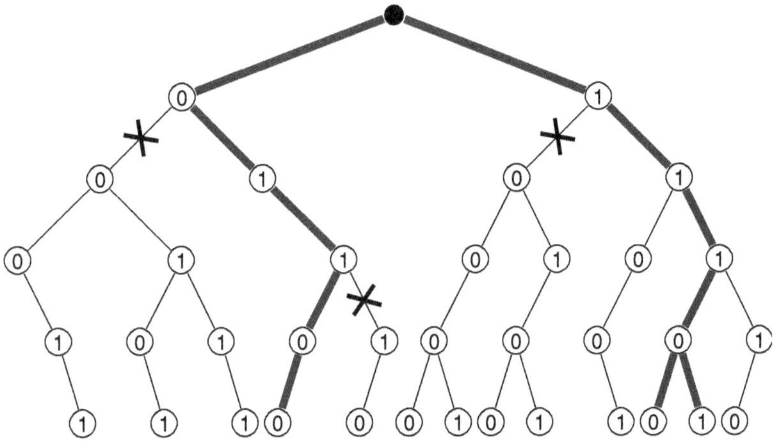

Figure 5.15 Subtree (in bold) corresponding to $N^{(e)} = 2$ in Example 5.6.

As expected for small $N_t^{(e)}$ $(1 \leq N_t^{(e)} \leq 10)$, the relative error RE of the estimator $|\widetilde{\mathcal{X}^*}|$ is large and it underestimates $|\mathcal{X}^*|$. Starting from $N_t^{(e)} = 25$ the estimator stabilizes. Note that for $N_t^{(e)} = 100$ the estimator is exact because $|\mathcal{X}^*| = 100$. Recall that the optimal zero variance SIS pdf over the $n + 1$ paths

$$(00\cdots00),\ (10\cdots00),\ (11\cdots00), \ldots,\ (11\cdots10),\quad (11\cdots11)$$

is $g^*(\boldsymbol{x}) = 1/(n + 1)$.

ii. Counting the number of $s - t$ paths in a graph $G = G(V, E)$ that has $m - 1$ intermediate vertices, such that there is exactly one $s - t$ path of length k,

Table 5.2 The Efficiencies of the SE Algorithm 5.3 for the 2-SAT Model with $|\mathcal{X}^*| = 100$

$(N^{(e)},\quad M)$	$\widetilde{|\mathcal{X}^*|}$	RE
$(N^{(e)} = 1,\ M = 500)$	11.110	0.296
$(N^{(e)} = 5,\ M = 100)$	38.962	0.215
$(N^{(e)} = 10,\ M = 50)$	69.854	0.175
$(N^{(e)} = 25,\ M = 20)$	102.75	0.079
$(N^{(e)} = 50,\ M = 10)$	101.11	0.032
$(N^{(e)} = 75,\ M = 5)$	100.45	0.012
$(N^{(e)} = 100,\ M = 1)$	100.00	0.000

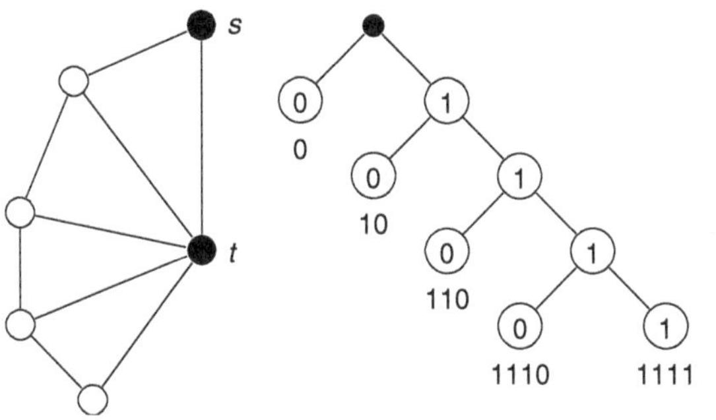

Figure 5.16 A $s - t$ graph and its associated tree with $|\mathcal{X}^*| = 5$ paths.

for each $k = 1, \ldots, m$. Thus $|\mathcal{X}^*| = m$. Figure 5.16 presents such a graph with $m = 5$, and its associated tree of solutions. The results for the graph with $|\mathcal{X}^*| = 100$ paths were similar as in Table 5.2 for the 2-SAT example above.

5.4 APPLICATIONS OF SE

Below we present several possible applications of SE. As usual we assume that there exists an associated polynomial time decision-making oracle.

5.4.1 Counting the Number of Trajectories in a Network

We show how to use SE for counting the number of trajectories (paths) in a network with a fixed source and sink. We demonstrate this for $N^{(e)} = 1$. The modification to $N^{(e)} > 1$ is straightforward. Consider the following two toy examples.

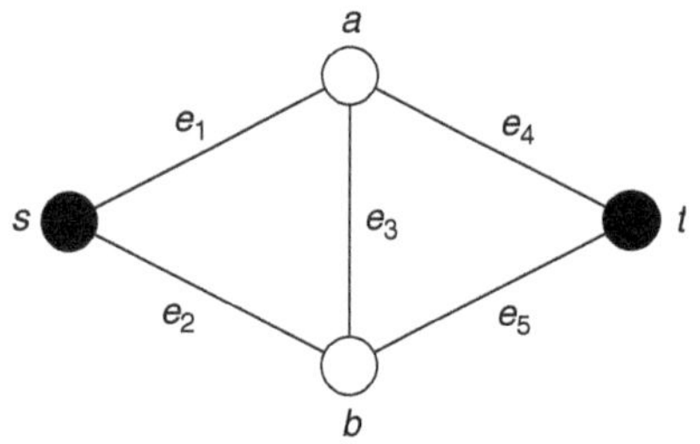

Figure 5.17 Bridge network: count the number of paths from s to t.

EXAMPLE 5.7 ***Bridge Network***

Consider the undirected graph in Figure 5.17. Suppose we wish to count the four trajectories

$$(e_1, e_4),\ (e_1, e_3, e_5),\ (e_2, e_3, e_4),\ (e_2, e_5) \tag{5.11}$$

between the source node s and sink node t.

1. **Iteration 1** Starting from s we have two nodes, e_1 and e_2, and the associated vector (x_1, x_2). Because each x is binary, we have the following four combination: (00), (01), (10), (11). Note that only the trajectories (01), (10) are relevant here because (00) can not be extended to node t, while the trajectory (11) is redundant given (01), (10). We have thus $N_1 = 2$ and $\nu_1 = 2$.
2. **Iteration 2** Assume that we selected randomly the path (01) among the two, (01) and (10). By doing so we arrive at the node b containing the edges (e_2, e_3, e_5). According to SE only the edges e_3 and e_5 are relevant. As before, their possible combinations are (00), (01), (10), (11). Arguing similarly to **Iteration 1** we have that $N_2 = 2$ and $\nu_2 = 2$. Consider separately the two possibilities associated with edges e_5 and e_3.
(i) Edge e_5 is selected. In this case, we can go directly to the sink node t and thus deliver an exact estimator $|\widehat{\mathcal{X}^*}| = \nu_1 \nu_2 = 2 \cdot 2 = 4$. The resulting path is (e_2, e_5).
3. **Iteration 3**
(ii) Edge e_3 is selected. In this case, we go to t via the edge e_4. It is readily seen that $N_3 = 1,\ \nu_3 = 1$. We have again $|\widehat{\mathcal{X}^*}| = \nu_1 \nu_2 \nu_3 = 2 \cdot 2 \cdot 1 = 4$. The resulting path is (e_2, e_3, e_4).

Note that if we select the combination (10) instead of (01) we would get again $N_2 = 2$ and $\nu_2 = 2$, and thus again an exact estimator $|\widehat{\mathcal{X}^*}| = 4$.

If instead of (5.11) we would have a directed graph with the following three trajectories

$$(e_1, e_4), \quad (e_1, e_3, e_5), \quad (e_2, e_5), \tag{5.12}$$

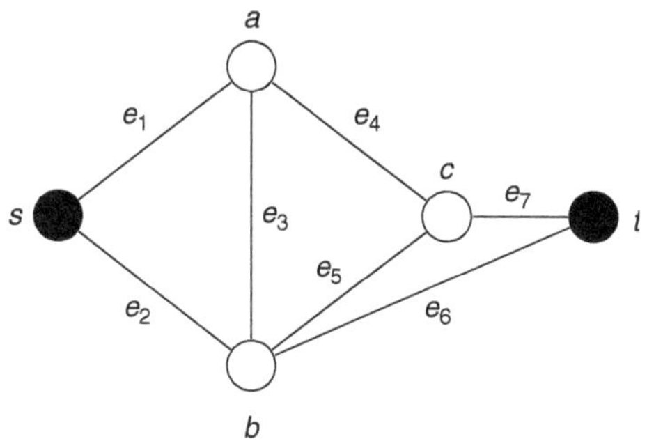

Figure 5.18 Extended bridge network: number of paths from s to t.

then we obtain (with probability 1/2) an estimator $\widehat{|\mathcal{X}^*|} = 2$ for the path (e_2, e_5) and (with probability 1/4) an estimator $\widehat{|\mathcal{X}^*|} = 4$ for the paths (e_1, e_4) and (e_1, e_3, e_5), respectively. □

EXAMPLE 5.8 ***Extended Bridge***

Figure 5.18 presents an extended version of the bridge network in Figure 5.17. We have the following seven trajectories:

$$(e_1, e_4, e_7),\ (e_1, e_3, e_6),\ (e_1, e_3, e_5, e_7),\ (e_1, e_4, e_5, e_6),$$
$$(e_2, e_6)\quad (e_2, e_5, e_7),\ (e_2, e_3, e_4, e_7). \tag{5.13}$$

1. **Iteration 1** This iteration coincides with **Iteration 1** of Example 5.7. We have $N_1 = 2$ and $\nu_1 = 2$.
2. **Iteration 2** Assume that we selected randomly the combination (01) from the two, (01) and (10). By doing so we arrive at node b containing the edges (e_2, e_3, e_5, e_6). According to SE only (e_3, e_5, e_6) are relevant. Of the seven combinations, only (001), (010), (100) are relevant; there is no path through (000), and the remaining ones are redundant because they result in the same trajectories as the above three. Thus, we have $N_2 = 3$ and $\nu_2 = 3$. Consider separately the three possibilities associated with edges e_6, e_5 and e_3.
(i) Edge e_6 is selected. In this case, we can go directly to the sink node t and thus deliver $\widehat{|\mathcal{X}^*|} = \nu_1\nu_2 = 2 \cdot 3 = 6$. The resulting path is (e_2, e_6). Note that if we select either e_5 or e_3, there is no direct access to the sink t.
3. **Iteration 3**
(ii) Edge e_5 is selected. In this case we go to t via the edge e_7. It is readily seen that $N_3 = 1$, $\nu_3 = 1$. We have again $\widehat{|\mathcal{X}^*|} = \nu_1\nu_2\nu_3 = 2 \cdot 3 \cdot 1 = 6$. The resulting path is (e_2, e_5, e_7).

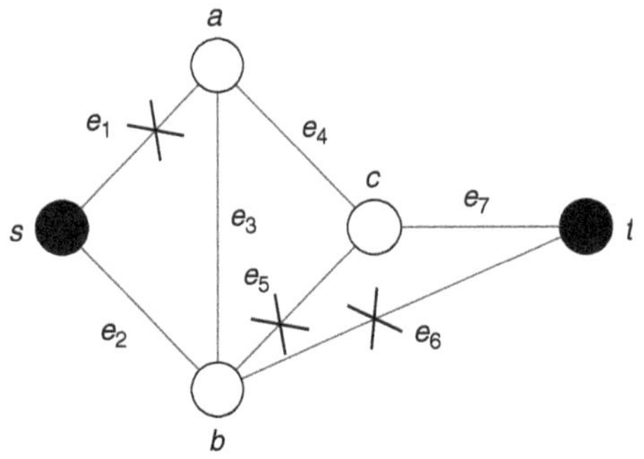

Figure 5.19 Subtree consisting of the path (e_2, e_3, e_4, e_7).

(iii) Edge e_3 is selected. By doing so we arrive at node a (the intersection of (e_1, e_3, e_4)). The only relevant edge is e_4. We have $N_3 = 1$, $\nu_3 = 1$.

4. **Iteration 4** We proceed with the path (e_2, e_3, e_4), which arrived to point c, the intersection of (e_4, e_5, e_7). The only relevant edge among (e_4, e_5, e_7) is e_7. We have $N_4 = 1$, $\nu_4 = 1$ and $|\widehat{\mathcal{X}^*}| = \nu_1 \nu_2 \nu_3 \nu_4 = 2 \cdot 3 \cdot 1 \cdot 1 = 6$. The resulting path is (e_2, e_3, e_4, e_7).

Figure 5.19 presents the subtree corresponding to the path (e_2, e_3, e_4, e_7) for the extended bridge in Figure 5.18.

It can be seen that if we choose the combination (10) instead of (01) we obtain $|\widehat{\mathcal{X}^*}| = 8$ for all four remaining cases. Below we summarize all seven cases.

Path	Probability	Estimator
(e_1, e_4, e_7)	$1/2 \cdot 1/2 \cdot 1/2$	8
(e_1, e_3, e_6)	$1/2 \cdot 1/2 \cdot 1/2$	8
(e_1, e_3, e_5, e_7)	$1/2 \cdot 1/2 \cdot 1/2 \cdot 1$	8
(e_1, e_4, e_5, e_6)	$1/2 \cdot 1/2 \cdot 1/2 \cdot 1$	8
(e_2, e_6)	$1/2 \cdot 1/3$	6
(e_2, e_5, e_7)	$1/2 \cdot 1/3 \cdot 1$	6
(e_2, e_3, e_4, e_7)	$1/2 \cdot 1/3 \cdot 1 \cdot 1$	6

Averaging over the all cases we obtain $|\widehat{\mathcal{X}^*}| = 4 \cdot 1/8 \cdot 8 + 3 \cdot 1/6 \cdot 6 = 7$ and thus, the exact value. □

Consider now the case $N^{(e)} > 1$. In particular, consider the graph in Figure 5.18 and assume that $N^{(e)} = 2$. In this case, at iteration 1 of SE Algorithm 5.3 both edges e_1 and e_2 will be selected. At iteration 2 we have to choose randomly two nodes out of the four: e_3, e_4, e_5, e_6. Assume that e_4 and e_5 are selected. Note that by selecting e_5, we completed an entire path, se_2e_6t. Note, however, that the second path that goes through the edges e_3, e_4 will be not yet completed, since finally it can be either $se_3e_4e_7t$ or $se_3e_4e_5e_6t$. In both cases, the shorter path se_2e_6t must be synchronized (length-wise) with the longer ones $se_3e_4e_7t$ or $se_3e_4e_5e_6t$ in the sense that, depending on whether $se_3e_4e_5e_6t$ or $se_3e_4e_7t$ is selected, we have to insert into se_2e_6t either one auxiliary edge from e_6 to t (denoted $e_6e_6^{(1)}$) or two auxiliary ones from e_6 to t (denoted by $e_6e_6^{(1)}$ and $e_6^{(1)}e_6^{(2)}$). The resulting path (with auxiliary edges) will be either $se_2e_6e_6^{(1)}t$ or $se_2e_6e_6^{(1)}e_6^{(2)}t$.

It follows from above that while adopting Algorithm 5.3 with $N^{(e)} > 1$ for counting the number of paths in a general network, one will have to synchronize on-line all its paths with the longest one by adding some auxiliary edges until all paths will match length-wise with the longest one.

5.4.2 SE for Probabilities Estimation

Algorithm 5.3 can be modified easily for the estimation of different probabilities in a network, such as the probability that the length $S(\boldsymbol{X})$ of a randomly chosen path $\boldsymbol{X}$ is greater than a fixed number γ, that is,

$$\ell = \mathbb{P}(S(\boldsymbol{X}) \geq \gamma).$$

Assume first that the length of each edge equals unity. Then the length $S(\boldsymbol{x})$ of a particular path $\boldsymbol{x}$ equals the number of edges from s to t on that path. The corresponding number of iterations is $(S(\boldsymbol{x}) - n_0)/r$ and the corresponding probability is

$$\ell = \mathbb{P}((S(\boldsymbol{X}) - n_0)/r \geq \gamma) = \mathbb{E}[I_{\{(S(\boldsymbol{X})-n_0)/r \geq \gamma\}}].$$

Clearly, there is no need to calculate the weights ν_t and the length $S(\boldsymbol{x}_j)$ of each path $\boldsymbol{x}_j$, $j = 1, \ldots, N^{(e)}$.

In cases where the lengths of the edges are different from one, $S(\boldsymbol{x})$ represents the sum of the lengths of edges associated with that path $\boldsymbol{x}$.

Algorithm 5.4 ***SE for Estimation of Probabilities***

1. ***Iteration 1***
   - ***Full Enumeration.*** *Same as in Algorithm 5.3.*
   - ***Calculation of the First Weight Factor.*** *Redundant. Instead, store the lengths of the corresponding edges* $\eta_{1,1}, \ldots, \eta_{1,N^{(e)}}$.
2. ***Iteration*** $t, t \geq 2$
   - ***Full Enumeration.*** *Same as in Algorithm 5.3.*
   - ***Stochastic Enumeration.*** *Same as in Algorithm 5.3.*
   - ***Calculation of the*** t***-th Weight Factor.*** *Redundant. Instead, store the lengths of the corresponding edges* $\eta_{t,1}, \ldots, \eta_{t,N^{(e)}}$.
3. ***Stopping Rule***
   - *Proceed with iteration* t, $t = 1, \ldots, (n - n_0)/r$ *and compute*

$$I_j = I_{\{(S(X_j)-n_0)/r \geq \gamma\}}, \quad j = 1, \ldots, N^{(e)}, \tag{5.14}$$

where, as before, $S(\boldsymbol{x})$ is the length of path $\boldsymbol{x}$ representing the sum of the lengths of the edges associated with $\boldsymbol{x}$.

4. ***Final Estimator***
   - *Run Algorithm 5.4 for M independent replications and deliver*

$$\widetilde{\ell} = \frac{1}{MN^{(e)}} \sum_{k=1}^{M} \sum_{j=1}^{N^{(e)}} I_{\{(S(\boldsymbol{X}_{jk})-n_0)/r \geq \gamma\}} \tag{5.15}$$

as an unbiased estimator of ℓ.

We performed various numerical studies with Algorithm 5.4 and found that it performs properly, provided that γ is chosen such that ℓ is not a rare-event probability; otherwise, one needs to use the importance sample method. For example, for a random Erdös-Rényi graph with 15 nodes and 50 edges (see Remark 5.3 below) we obtained via full enumeration that the number of valid paths is 643,085. Furthermore, the probability $\ell = \mathbb{P}(S(\boldsymbol{X}) \geq \gamma)$ that the length $S(\boldsymbol{X})$ of a randomly chosen path $\boldsymbol{X}$ is greater than $\gamma = 10$ equals $\ell = 0.8748$. From our numerical results with $N^{(e)} = 100$ and $M = 100$ based on 10 runs we obtained with Algorithm 5.4 an estimate $\widehat{\ell} = 0.8703$ with relative error about 1%.

Remark 5.3 ***Erdös-Rényi Random Graphs Generation***

Random graphs generation is associated with Paul Erdös and Alfred Rényi. We consider indirected graphs generated according to what is called the $G(n, p)$ Erdös-Rényi random model [44].

In the $G(n, p)$ graph (n is fixed), the graph is constructed by connecting nodes randomly. Each edge is included in the graph with probability p independent from every other edge. Equivalently, all graphs with n nodes and m edges have the same probability

$$p^m(1-p)^{\binom{n}{2}-m}.$$

A simple way to generate a random graph in $G(n, p)$ is to consider each of the possible $\binom{n}{2}$ edges in some order and then independently add each edge to the graph with probability p. Note that the expected number of edges in $G(n, p)$ is $p\binom{n}{2}$, and each vertex has expected degree $p(n-1)$. Clearly as p increases from 0 to 1, the model becomes more dense in the sense that is it is more likely that it will include graphs with more edges than less edges.

5.4.3 Counting the Number of Perfect Matchings in a Graph

Here we deal with application of SE to compute the number of matchings in a graph with particular emphasis on the number of *perfect matchings* in a balanced bipartite graph.

Recall that a graph $G = G(V, E)$ is bipartite if it has no circuits of odd length. It has the following property: the set of vertices V can be partitioned in two disjoint sets, $V_1 = (v_{11}, \dots, v_{1n_1})$ and $V_2 = (v_{21}, \dots, v_{2n_2})$, such that each edge in E has one vertex in V_1 and one in V_2. Thus, we denote a bipartite graph by $G = G(V_1, V_2, E)$. In a balanced bipartite graph, the two subsets have the same cardinality: $|V_1| = |V_2| = n$. A matching of a graph $G = G(V, E)$ is a subset of the edges with the property that no two edges share the same node. In other words, a matching is a collection of edges $M \subseteq E$ such that each vertex occurs at most once in M. A perfect matching is a matching that matches all vertices. Thus, a perfect matching in a balanced bipartite graph has size n.

Let $G = G(V_1, V_2, E)$ be a balanced bipartite graph. Its associated biadjacency matrix $\boldsymbol{A}$ is an $n \times n$ binary matrix $\boldsymbol{A}$ defined by

$$a_{ij} = \begin{cases} 1 & \text{if } \left(v_{1i}, v_{2j}\right) \in E; \\ 0 & \text{otherwise.} \end{cases}$$

It is well known [90] that the number of perfect matchings in a balanced bipartite graph coincides with the permanent of its associated biadjacency matrix $\boldsymbol{A}$. The permanent of a binary $n \times n$ matrix $\boldsymbol{A}$ is defined as

$$\text{per}(\boldsymbol{A}) = \sum_{\sigma \in \Sigma_n} \prod_{i=1}^{n} a_{i\sigma_i}, \tag{5.16}$$

where Σ_n is the set of all permutations $\sigma = (\sigma_1, \dots, \sigma_n)$ of $(1, \dots, n)$.

EXAMPLE 5.9

Consider the bipartite graph shown in Figure 5.20. It has the following three perfect matchings:

$$\begin{aligned} M_1 &= [(a_1, b_1), (a_2, b_2), (a_3, b_3)], \\ M_2 &= [(a_1, b_3), (a_2, b_1), (a_3, b_2)], \\ M_3 &= [(a_1, b_3), (a_2, b_2), (a_3, b_1)]. \end{aligned} \tag{5.17}$$

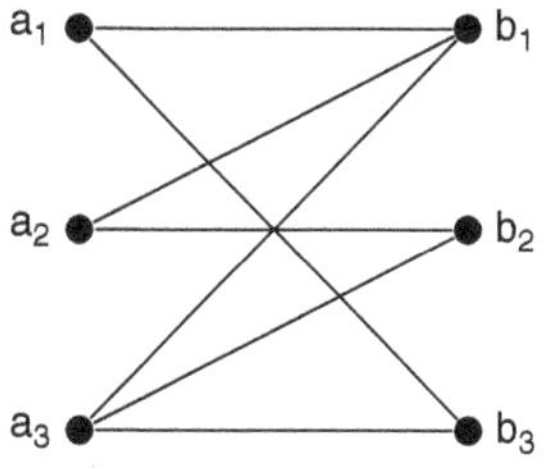

Figure 5.20 The bipartite graph.

Its associated biadjacency matrix is

$$A = \begin{pmatrix} 1 & 0 & 1 \\ 1 & 1 & 0 \\ 1 & 1 & 1 \end{pmatrix}. \tag{5.18}$$

The permanent of A is

$$a_{11}a_{22}a_{33} + a_{13}a_{21}a_{32} + a_{13}a_{22}a_{31} + \text{zero terms} = 3.$$

We will show how SE works for $N^{(e)} = 1$. Its extension to $N^{(e)} > 1$ is simple. We say that an edge is active if the outcome of the corresponding variable is 1 and passive otherwise.

a. Iteration 1 Let us start from node a_1. Its degree is 2, and the corresponding edges are (a_1, b_1) and (a_1, b_3). To each of these edges we associate a Ber($p = 1/2$) random variable. An outcome of 1 (or 0) means that the associated edge is active (or passive). We let these two Bernoulli variables be independent. Thus, the possible outcomes are (00), (01), (10), (11). For instance, (10) means that edge (a_1, b_1) is active and (a_1, b_3) is passive, while (01) means it is the other way around. However, only (01) and (10) are relevant in as much as neïther (00) nor (11) define a perfect matching. Employing the oracle, we obtain that each of the combinations (01) and (10) is valid, and starting from (a_1, b_1) and (a_1, b_3), we generate two different perfect matchings (see (5.17)); we therefore have that $N_1 = 2$, $\nu_1 = 2$.

We next proceed separately with the outcomes (10) and (01).

- **Outcome** (10)

  b. Iteration 2 Recall that the outcome (10) means that (a_1, b_1) is active. This automatically implies that all three neighboring edges, (a_1, b_3), (a_3, b_1), (a_2, b_1), must be passive. Using the perfect matching oracle we will arrive at the next active edge, which is (a_2, b_2). Because the degree of node a_2 is two and since (a_2, b_1) must be passive, we have that $N_2 = 1$, $\nu_2 = 1$.

  c. Iteration 3 Because (a_2, b_2) is active, (a_3, b_2) must be passive. The degree of node a_3 is three, but since (a_3, b_2) and (a_3, b_1) are passive, (a_3, b_3) must be the only available active edge. This implies that $N_3 = 1$, $\nu_3 = 1$. The resulting estimator of $|\mathcal{X}^*|$ is $\widehat{|\mathcal{X}^*|} = 2 \cdot 1 \cdot 1 = 2$.

- **Outcome** (01)

  b. Iteration 2 Because (01) means that (a_1, b_3) is active, we automatically set the neighboring edges (a_1, b_1), (a_3, b_3) as passive. Using an oracle we shall arrive at node a_3, which has degree three. As (a_1, b_1) is passive, it is easily seen that each edge, (a_3, b_1) and (a_3, b_2), will become active with probability 1/2. This means that with (a_3, b_1) and (a_3, b_2) we are in a similar situation (in the sense of active-passive edges) to that of (a_1, b_1), and (a_1, b_3). We thus have for each case $N_2 = 2$, $\nu_2 = 2$.

c. Iteration 3 It is readily seen that both cases (a_3, b_1), and (a_3, b_2) lead to $N_3 = 1$, $\nu_3 = 1$. The resulting estimator of $|\mathcal{X}^*|$ is $\widehat{|\mathcal{X}^*|} = 2 \cdot 2 \cdot 1 = 4$. Because each initial edge (a_1, b_1) and (a_1, b_3) at **Iteration 1** is chosen with probability 1/2, by averaging over both cases we obtain the exact result, that is, $\mathbb{E}[\widehat{|\mathcal{X}^*|}] = |\mathcal{X}^*| = 3$.

It is not difficult to see that when we select $N^{(e)} = 2$ instead of $N^{(e)} = 1$ we obtain $\widehat{|\mathcal{X}^*|} = 3/2 \cdot 2 \cdot 1 = 3$, that is, the exact value $|\mathcal{X}^*| = 3$.

Another view is to consider the equivalence with the permanent. The sample space might be defined to be all permutations $\boldsymbol{x} \in \Sigma_3$, where $x_i \in \{1, 2, 3\}$ represents using edge $(i \in V_1, x_i \in V_2)$ in the matching. The solution set consists of the three permutations representing (5.17):

$$\mathcal{X} = \{(123), (312), (321)\}.$$

The optimal zero variance importance sampling pdf is uniform on $\mathcal{X}$:

$$g^*(\boldsymbol{x}) = \begin{cases} 1/3 & \text{for } \boldsymbol{x} \in \mathcal{X}; \\ 0 & \text{otherwise.} \end{cases}$$

The SE procedure above with $N^{(e)} = 1$ leads to an importance sampling pdf g satisfying

$$g(\boldsymbol{x}) = \begin{cases} 1/2 & \text{for } \boldsymbol{x} = (123); \\ 1/4 & \text{for } \boldsymbol{x} = (312); \\ 1/4 & \text{for } \boldsymbol{x} = (321); \\ 0 & \text{otherwise.} \end{cases}$$

The associated SE estimator satisfies

$$\mathbb{P}_g(\widehat{|\mathcal{X}^*|} = 2) = \mathbb{P}_g(\widehat{|\mathcal{X}^*|} = 4) = \frac{1}{2},$$

yielding $\mathbb{E}_g[\widehat{|\mathcal{X}^*|}] = 3$, and $\mathbb{V}\mathrm{ar}_g[\widehat{|\mathcal{X}^*|}] = (1/2) \times (4 + 16) - 9 = 1$. □

5.4.4 Counting SAT

Note that, although for general SAT the decision-making is an NP-hard problem, there are several very fast heuristics for this purpose. The most popular one is the famous DPLL solver [35] (see Remark 5.4 below), on which two main heuristic algorithms are based for approximately counting with emphasis on SAT. The first, called `ApproxCount` and introduced by Wei and Selman in [126], is a local search method that uses Markov chain Monte Carlo sampling to approximate the true counting quantity. It is fast and has been shown to yield good estimates for feasible solution counts, but there are no guarantees as to uniformity of the Markov chain Monte Carlo samples. Gogate and Dechter [56, 57] recently proposed an alternative counting technique called `SampleMinisat`, based on sampling from the so-called backtrack-free search space of a Boolean formula through `SampleSearch`. An approximation of the search tree thus found is used as the importance sampling density instead of a uniform distribution over all solutions. They also derived a lower bound for the unknown counting quantity. Their empirical studies demonstrate the superiority of their method over its competitors.

Remark 5.4 ***DPLL Algorithm***

The Davis-Putman-Logemann-Loveland (DPLL) algorithm [35] is a complete, backtracking-based algorithm for deciding the satisfiability of propositional logic formulae in conjunctive normal form, that is for solving the CNF-SAT problem. DPLL is a highly efficient procedure and after more than 40 years still forms the basis for most efficient complete SAT solvers, as well as for many theorem provers for fragments of first-order logic.

The basic backtracking algorithm runs by choosing a literal, assigning a truth value to it, simplifying the formula, and then recursively checking whether the simplified formula is satisfiable. If this is the case, the original formula is satisfiable; otherwise, the same recursive check is done assuming the opposite truth value. This is known as the splitting rule, as it splits the problem into two simpler subproblems. The simplification step essentially removes all clauses that become true under the assignment from the formula and all literals that become false from the remaining clauses.

5.5 NUMERICAL RESULTS

Here we present numerical results with the multiple-OSLA and SE algorithms. In particular, we use the multiple-OSLA Algorithm 5.2 for counting SAWs. The reason for doing so is that we are not aware of any polynomial decision-making algorithm (oracle) for SAWs. For the remaining problems we use the SE Algorithm 5.3 because polynomial decision-making algorithms (oracles) are available for them. If not stated otherwise, we use $r = 1$.

To achieve high efficiency of the SE Algorithm 5.3, we set M as suggested in Section 5.3. In particular,

1. For SAT models to set $M = 1$. Note that by doing so, $N^{(e)} = K$, where K is the allowed budget. Also, in this case $N^{(e)}$ is the only parameter of SE. For the instances where we occasionally obtained (after simulation) an exact $|\mathcal{X}^*|$, that is where $|\mathcal{X}^*|$ is relatively small and where we originally set $N^{(e)} \geq |\mathcal{X}^*|$, we purposely reset $N^{(e)}$ (to be fair to the other methods) by choosing a new $N_*^{(e)}$, satisfying $N_*^{(e)} < |\mathcal{X}^*|$ and run SE again. By doing so we prevent SE from being an ideal (zero variance) estimator.
2. For other counting problems, such as counting the number of perfect matchings (permanent) and the number of paths in a network, to perform small pilot runes with several values of M and select the best one.

We use the following notations:

1. $N_t^{(e)}$ denotes the number of elites at iteration t.
2. n_t denotes the level reached at iteration t.
3. $\nu_t = N_t / N_t^{(e)}$ denotes the weight factor at iteration t.

5.5.1 Counting SAW

Tables 5.3 and 5.4 present the performance of the multiple-OSLA Algorithm 5.2 for SAW for $n = 500$ and $n = 1,000$, respectively, with $r = 2$, $n_0 = 4$, and $M = 20$ (see (5.8)). This corresponds to the initial values $N_0^{(e)} = 100$ and $N_1 = 780$ (see also iteration $t = 0$ in Table 5.5). Based on the runs in Table 5.3, we found RE $= 0.0685$. Based on the runs in Table 5.4, we found RE $= 0.0901$. Table 5.5 presents dynamics of one of the runs of the multiple-OSLA Algorithm 5.2 for $n = 500$.

5.5.2 Counting the Number of Trajectories in a Network

5.5.2.1 *Model 1: From Roberts and Kroese [96] with n = 24 Nodes*

Table 5.6 presents the performance of the SE Algorithm 5.3 for **Model 1** taken from Roberts and Kroese [96] with the following adjacency (24×24) matrix:

$$
\begin{pmatrix}
0&0&0&1&0&0&0&0&0&0&0&1&0&0&1&1&1&0&1&1&0&0&0&0\\
0&0&0&0&0&0&0&0&0&0&1&0&1&1&1&0&0&0&0&0&0&1&0&0\\
0&0&0&0&1&0&1&0&0&0&0&0&1&0&0&0&0&0&0&0&0&0&0&0\\
1&0&0&0&0&0&1&0&0&0&0&1&0&0&0&0&1&0&1&0&0&0&0&0\\
0&0&1&0&0&0&0&0&0&0&0&0&0&0&0&1&1&0&1&0&0&0&0&0\\
0&0&0&0&0&0&1&0&0&0&1&0&1&0&1&0&0&0&0&0&0&0&0&0\\
0&0&1&1&0&1&0&0&1&0&0&1&1&0&0&0&1&0&0&0&0&0&0&0\\
0&0&0&0&0&0&0&0&1&1&0&0&0&1&0&0&1&0&0&0&0&0&0&0\\
0&0&0&0&0&0&1&1&0&0&0&0&0&0&0&1&1&0&0&0&0&1&0&0\\
0&0&0&0&0&0&0&1&0&0&0&0&0&0&1&0&0&0&0&0&0&0&0&0\\
0&1&0&0&0&1&0&0&0&0&0&0&0&0&0&1&1&0&0&1&1&0&0&0\\
1&0&0&1&0&0&1&0&0&0&0&0&1&0&0&0&0&0&0&0&0&0&1&0\\
0&1&1&0&0&1&1&0&0&0&0&1&0&0&0&0&0&0&1&0&1&0&1&0\\
0&1&0&0&0&0&0&1&0&0&0&0&0&0&0&0&0&0&1&0&0&0&0&0\\
1&1&0&0&0&1&0&0&0&1&0&0&0&0&0&0&0&0&0&0&0&0&0&0\\
1&0&0&0&1&0&0&0&1&0&1&0&0&0&0&0&0&0&0&1&0&0&1&1\\
1&0&0&1&1&0&1&1&1&0&1&0&0&0&0&0&0&0&1&1&0&0&1&0\\
0&0&0&0&0&0&0&0&0&0&0&0&0&0&0&0&0&0&0&0&0&1&1&0\\
1&0&0&1&1&0&0&0&0&0&0&0&1&1&0&0&1&0&0&0&0&0&0&0\\
1&0&0&0&0&0&0&0&0&0&1&0&0&0&0&1&1&0&0&0&0&0&1&0\\
0&0&0&0&0&0&0&0&0&0&1&0&1&0&0&0&0&0&0&0&0&0&0&0\\
0&1&0&0&0&0&0&0&1&0&0&0&0&0&0&0&0&1&0&0&0&0&0&0\\
0&0&0&0&0&0&0&0&0&0&0&1&1&0&0&1&1&1&0&1&0&0&0&1\\
0&0&0&0&0&0&0&0&0&0&0&0&0&0&0&1&0&0&0&0&0&0&1&0
\end{pmatrix}
$$

We set $N_t^{(e)} = 50$ and $M = 400$ to get a comparable running time.

Table 5.3 Performance of the multiple-OSLA Algorithm 5.2 for counting SAW for $n = 500$

Run	Iterations	$\widehat{|\mathcal{X}^*|}$	CPU (sec.)
1	248	4.799E+211	130.55
2	248	4.731E+211	130.49
3	248	4.462E+211	132.36
4	248	4.302E+211	136.22
5	248	5.025E+211	132.19
6	248	5.032E+211	131.79
7	248	4.397E+211	132.18
8	248	4.102E+211	131.60
9	248	4.820E+211	131.98
10	248	4.258E+211	131.73
Average	248	4.593E+211	132.11

Table 5.4 Performance of the multiple-OSLA Algorithm 5.2 for counting SAW for $n = 1,000$

Run	Iterations	$\widehat{|\mathcal{X}^*|}$	CPU (sec.)
1	497	2.514E+422	4008
2	497	2.629E+422	3992
3	497	2.757E+422	3980
4	497	2.354E+422	3975
5	497	2.200E+422	3991
6	497	2.113E+422	3991
7	497	2.081E+422	3970
8	497	2.281E+422	3983
9	497	2.504E+422	3982
10	497	2.552E+422	3975
Average	497	2.399E+422	3985

Based on the runs in Table 5.6, we found RE $= 0.0264$. Comparing the results of Table 5.6 with those in [96], it follows that the former is about 1.5 times faster than the latter.

5.5.2.2 *Model 2: Large Model (n = 200 vertices and 199 edges)*

Table 5.7 presents the performance of SE Algorithm 5.3 for **Model 2** with $N_t^{(e)} = 100$ and $M = 500$. We found for this model RE $= 0.03606$.

We also counted the number of paths for Erdös-Rényi random graphs with $p = \ln\ n/n$ (see Remark 5.3). We found that SE performs reliably (RE ≤ 0.05) for $n \leq 200$, provided the CPU time is limited to 5–15 minutes.

Table 5.5 Dynamics of a run of the multiple-OSLA Algorithm 5.2 for counting SAW for $n = 500$

t	n_t	$N_t^{(e)}$	N_t	$\widehat{\nu}_t$	$\widehat{\lvert \mathcal{X}_t^* \rvert}$
0	4	100	100	1.00	100
1	6	100	780	7.80	780
2	8	100	759	7.59	5.920E+03
3	10	100	746	7.46	4.416E+04
4	12	100	731	7.31	3.228E+05
5	14	100	733	7.33	2.366E+06
50	104	100	699	6.99	4.528E+44
100	204	100	695	6.95	7.347E+86
150	304	100	699	6.99	1.266E+129
200	404	100	694	6.94	1.809E+171
244	492	100	693	6.93	2.027E+208
245	494	100	693	6.93	1.405E+209
246	496	100	696	6.96	9.780E+209
247	498	100	701	7.01	6.856E+210
248	500	100	700	7.00	4.799E+211

Table 5.6 Performance of SE Algorithm 5.3 for counting trajectories in **Model 1** Graph with $N_t^{(e)} = 50$ and $M = 400$

Run	Iterations	$\widehat{\lvert \mathcal{X}^* \rvert}$	CPU
1	18.87	1.81E+06	4.218
2	18.83	1.93E+06	4.187
3	18.83	1.98E+06	4.237
4	18.83	1.82E+06	4.232
5	18.79	1.90E+06	4.225
6	18.83	1.94E+06	4.231
7	18.86	1.86E+06	4.207
8	18.77	1.87E+06	4.172
9	18.79	1.90E+06	4.289
10	18.81	1.92E+06	4.287
Average	18.82	1.89E+06	4.229

Table 5.8 presents the performance of SE Algorithm 5.3 for the Erdös-Rényi random graph with $n = 200$ using $N_t^{(e)} = 1$ and $M = 30,000$. We found RE = 2.04E−02.

Remark 5.5

The SE method has some limitations, in particular when dealing with nonsparse instances. In this case, one must use in SE the full enumeration step (employ the oracle) many times.

Table 5.7 Performance of SE Algorithm 5.3 for counting trajectories in **Model 2** with $N_t^{(e)} = 100$ and $M = 500$

Run	Iterations	$\widehat{|\mathcal{X}^*|}$	CPU
1	58.93	1.53E+07	157.37
2	58.86	1.62E+07	153.92
3	58.96	1.69E+07	153.90
4	59.09	1.65E+07	154.50
5	58.73	1.57E+07	153.38
6	58.96	1.57E+07	153.71
7	59.13	1.67E+07	153.91
8	58.43	1.51E+07	153.14
9	59.08	1.62E+07	153.87
10	58.90	1.59E+07	154.81
Average	58.91	1.60E+07	154.25

Table 5.8 Performance of SE Algorithm 5.3 for counting trajectories in the Erdös-Rényi random graph ($n = 200$) with $N_t^{(e)} = 1$ and $M = 30,000$

Run	Iterations	$\widehat{|\mathcal{X}^*|}$	CPU
1	80.08	1.43E+55	471.0
2	80.43	1.39E+55	499.0
3	80.51	1.41E+55	525.0
4	80.76	1.38E+55	507.9
5	80.43	1.45E+55	505.5
Average	80.44	1.41E+55	501.7

As a result, the CPU time of SE might increase dramatically. Consider, for example, the extreme case, a complete graph K_n. The exact number of s-t paths between any two vertices is given by

$$K(n) = \sum_{k=0}^{n-2} \frac{(n-2)!}{k!}.$$

Table 5.9 presents the performance of SE Algorithm 5.3 for K_{25} with $N_t^{(e)} = 50$ and $M = 100$. The exact solution is 7.0273E+022. For this model we found RE = 0.0248, so the results are still good. However, as n increases, the CPU time grows rapidly. For example, for K_{100} we found that using $N_t^{(e)} = 100$ and $M = 100$ the CPU time is about 5.1 hours.

5.5.3 Counting the Number of Perfect Matchings in a Graph

We present here numerical results for three models, one small and two large.

Table 5.9 Performance of SE Algorithm 5.3 for counting trajectories in K_{25} with $N_t^{(e)} = 50$ and $M = 100$

Run	Iterations	$\widehat{\|\mathcal{X}^*\|}$	CPU
1	23.00	7.07E+22	28.71
2	23.00	6.95E+22	28.99
3	23.00	7.10E+22	28.76
4	23.00	7.24E+22	28.33
5	23.00	7.03E+22	27.99
6	23.00	6.75E+22	28.32
7	23.00	7.18E+22	29.25
8	23.00	7.15E+22	29.51
9	23.00	6.85E+22	28.20
10	23.00	7.34E+22	28.12
Average	23.00	7.07E+22	28.62

5.5.3.1 *Model 1: (Small Model)*

Consider the following $\boldsymbol{A} = 30 \times 30$ matrix where the true number of perfect matchings (permanent) is $|\mathcal{X}^*| = 266$, obtained using full enumeration:

$$
\begin{pmatrix}
1&0&0&0&0&0&0&0&0&0&0&0&0&0&0&0&0&0&0&0&0&0&0&0&0&0&0&0&0&0\\
0&1&0&0&1&0&0&0&0&0&0&0&0&0&0&0&0&0&0&1&0&0&0&0&0&0&0&0&0&0\\
0&0&1&0&0&0&0&0&0&0&0&0&0&0&0&0&0&0&0&0&1&1&0&1&0&1&0&0&0&0\\
0&0&0&1&0&1&0&0&0&0&0&1&0&0&0&0&1&0&0&0&0&0&0&0&0&0&0&1&0&0\\
0&0&0&1&1&0&0&0&1&0&0&0&0&0&0&0&0&0&0&0&0&0&0&0&0&0&0&0&1&0\\
0&0&1&0&0&1&0&0&0&0&0&0&0&0&0&0&0&1&0&0&0&0&0&0&0&0&0&0&0&0\\
0&0&0&0&0&0&1&0&0&0&0&0&0&0&0&0&0&0&0&0&0&0&0&0&0&0&0&0&0&0\\
0&0&0&0&0&0&0&1&0&0&0&0&1&0&0&0&0&0&0&0&1&0&0&0&0&0&0&0&0&0\\
1&1&0&0&0&0&0&0&1&0&0&0&0&0&0&0&0&0&0&0&0&0&0&0&0&0&0&0&0&0\\
1&0&0&0&0&0&0&0&0&1&0&0&0&0&0&0&0&0&0&1&0&0&0&0&0&0&0&0&0&0\\
0&0&0&0&0&0&1&0&1&0&1&1&0&0&0&0&0&0&0&0&0&0&1&0&0&0&0&0&0&0\\
1&0&0&0&0&0&0&1&0&0&0&1&0&0&1&0&0&0&0&0&1&0&0&0&0&0&0&0&1&0\\
1&0&0&0&0&0&0&0&0&0&1&0&1&0&0&0&0&0&0&0&0&0&0&0&0&0&0&0&0&0\\
0&0&0&0&0&0&0&0&0&0&0&0&0&1&1&0&0&0&0&0&0&0&0&0&0&0&0&0&0&0\\
0&0&1&0&0&0&0&0&0&0&0&0&0&0&1&0&0&0&0&0&0&0&0&0&0&0&0&0&0&0\\
0&0&0&0&0&0&0&1&0&0&0&0&0&0&0&1&0&0&0&0&0&0&0&0&0&0&0&0&0&0\\
0&0&0&0&0&1&0&0&0&0&0&0&0&0&0&0&1&1&0&0&0&0&0&0&0&0&0&0&0&0\\
0&0&0&0&0&1&0&0&0&0&0&1&0&0&0&0&0&1&0&0&0&0&0&0&0&0&0&0&0&0\\
0&0&0&0&0&0&0&0&0&0&0&0&0&0&0&0&0&0&1&0&0&0&1&0&0&0&0&0&0&0\\
1&0&0&0&0&0&0&0&1&0&0&1&0&0&0&0&0&0&0&1&0&0&0&0&0&0&1&0&0&0\\
0&0&0&0&0&0&0&0&0&0&0&0&0&0&0&0&0&0&0&0&1&0&0&1&0&0&0&0&0&0\\
0&0&0&0&0&0&0&0&0&0&0&0&0&0&0&0&0&0&0&0&0&1&0&0&0&0&0&0&0&0\\
0&0&0&0&0&0&0&0&0&0&0&0&0&0&0&0&0&0&0&0&0&0&1&0&0&0&0&0&0&0\\
1&0&0&0&0&0&0&0&0&0&0&0&0&0&0&0&0&0&0&0&0&0&1&1&0&0&0&0&0&0\\
0&0&0&0&0&0&0&0&0&0&0&0&0&0&0&0&0&0&0&0&0&0&1&0&1&0&0&0&0&0\\
0&0&0&0&0&0&1&0&0&0&0&0&1&1&0&0&0&0&0&0&0&0&0&0&0&1&0&0&0&0\\
0&1&0&0&0&0&0&0&0&1&0&0&0&0&0&0&1&0&0&0&0&0&0&0&0&0&1&0&0&0\\
0&0&0&1&0&0&0&0&0&0&0&0&0&0&0&0&0&0&0&0&0&0&0&0&0&0&0&1&0&0\\
0&0&0&0&0&0&0&0&0&0&0&1&0&0&0&0&0&0&0&0&0&0&0&0&0&0&0&0&1&0\\
0&0&0&0&0&0&0&0&0&0&0&0&0&0&0&0&0&0&0&0&1&0&0&0&0&0&0&0&0&1
\end{pmatrix}
$$

Table 5.10 presents the performance of SE Algorithm 5.3 for ($N_t^{(e)} = 50$ and $M = 10$). We found that the relative error is 0.0268.

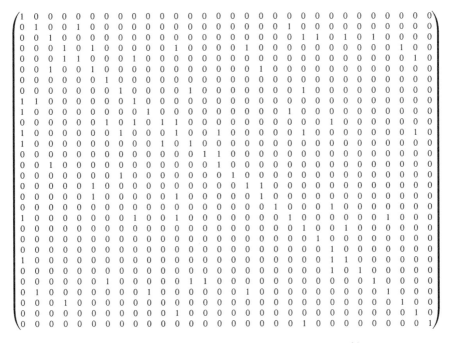

Table 5.10 Performance of SE Algorithm 5.3 for counting perfect matchings in **Model** 1 with $N_t^{(e)} = 50$ and $M = 10$

Run	Iterations	$\widehat{\lvert\mathcal{X}^*\rvert}$	CPU
1	24	264.21	2.056
2	24	269.23	2.038
3	24	270.16	2.041
4	24	268.33	2.055
5	24	272.10	2.064
6	24	259.81	2.034
7	24	271.62	2.035
8	24	269.47	2.050
9	24	264.86	2.059
10	24	273.77	2.048
Average	24	268.36	2.048

Table 5.11 Performance of SE Algorithm 5.3 for counting perfect matchings in **Model 2** with $N_t^{(e)} = 100$ and $M = 100$

Run	Iterations	$\widehat{\lvert\mathcal{X}^*\rvert}$	CPU
1	93	1.58E+05	472.4
2	93	1.77E+05	482.8
3	93	1.77E+05	472.4
4	93	1.65E+05	482.0
5	93	1.58E+05	475.7
6	93	1.78E+05	468.9
7	93	1.76E+05	469.3
8	93	1.73E+05	480.9
9	93	1.74E+05	473.7
10	93	1.78E+05	472.0
Average	93	1.71E+05	475.0

Applying the splitting algorithm for the same model using $N = 15,000$ and $\rho = 0.1$, we found that the relative error is 0.2711. It follows that SE is about 100 times faster than splitting.

5.5.3.2 *Model 2 with* 100 x 100 *(Large Model)*

Table 5.11 presents the performance of SE Algorithm 5.3 for **Model 2** matrix $N_t^{(e)} = 100$ and $M = 100$. The relative error is 0.0434.

Table 5.12 Performance of SE Algorithm 5.3 for counting perfect matchings in the Erdös-Rényi random graph using $N_t^{(e)} = 1$ and $M = 20,000$

Run	Iterations	$\widehat{\lvert \mathcal{X}^* \rvert}$	CPU
1	50	3.60E+07	3.29E+02
2	50	3.40E+07	3.50E+02
3	50	3.54E+07	3.30E+02
4	50	3.61E+07	3.40E+02
5	50	3.60E+07	3.33E+02
6	50	3.79E+07	3.27E+02
7	50	3.58E+07	3.33E+02
8	50	3.42E+07	3.35E+02
9	50	3.88E+07	3.25E+02
10	50	3.55E+07	3.19E+02
Average	50	3.60E+07	332.09179

5.5.3.3 Model 3: Erdös-Rényi Graph with $n = 100$ Edges and $p = 0.07$

Table 5.12 presents the performance of SE Algorithm 5.3 for the Erdös-Rényi graph with $n = 100$ edges and $p = 0.07$. We set $N_t^{(e)} = 1$ and $M = 20,000$. We found that the relative error is 0.0387.

We also applied the SE algorithm for counting general matchings in a bipartite graph. The quality of the results was similar to that for paths counting in a network.

5.5.4 Counting SAT

Here, we present numerical results for several SAT models. We set $r = 1$ in the SE algorithm for all models. Recall that for 2-SAT we can use a polynomial decision-making oracle [35], whereas for K-SAT ($K > 2$) heuristic based on the DPLL solver [35, 36] is used. Note that SE Algorithm 5.3 remains exactly the same when a polynomial decision-making oracle is replaced by a heuristic one, such as the DPLL solver. Running both SAT models we found that the CPU time for 2-SAT is about 1.3 times faster than for K-SAT ($K > 2$).

Table 5.13 presents the performance of SE Algorithm 5.3 for the 3-SAT 75×325 model with exactly 2258 solutions. We set $N_t^{(e)} = 20$ and $M = 100$. Based on those runs, we found RE = 0.0448.

Table 5.14 presents the performance of SE Algorithm 5.3 for the 3-SAT 75×270 model. We set $N_t^{(e)} = 100$ and $M = 100$. The exact solution for this instance is $|\mathcal{X}^*| = 1,346,963$ and is obtained via full enumeration using the backward method. It is interesting to note that, for this instance (with relatively small $|\mathcal{X}^*|$), the CPU time of the exact backward method is 332 seconds (compared with an

Table 5.13 Performance of SE Algorithm 5.3 for the 3-SAT 75 × 325 model

Run	Iterations	$\widehat{\lvert\mathcal{X}^*\rvert}$	CPU
1	75	2359.780	2.74
2	75	2389.660	2.77
3	75	2082.430	2.79
4	75	2157.850	2.85
5	75	2338.100	2.88
6	75	2238.940	2.75
7	75	2128.920	2.82
8	75	2313.390	3.04
9	75	2285.910	2.81
10	75	2175.790	2.85
Average	75	2247.077	2.83

Table 5.14 Performance of SE Algorithm 5.3 for the 3-SAT 75 × 270 model

Run	Iterations	$\widehat{\lvert\mathcal{X}^*\rvert}$	CPU
1	75	1.42E+06	16.88
2	75	1.37E+06	16.91
3	75	1.31E+06	17.24
4	75	1.35E+06	17.38
5	75	1.31E+06	16.75
6	75	1.32E+06	16.81
7	75	1.32E+06	16.51
8	75	1.25E+06	16.27
9	75	1.45E+06	16.3
10	75	1.33E+06	16.95
Average	75	1.35E+06	16.80

average 16.8 seconds time of SE in Table 5.14). Based on those runs, we found that RE = 0.0409.

Table 5.15 presents the performance of SE Algorithm 5.3 for the 3-SAT 300 × 1080 model. We set $N_t^{(e)} = 300$ and $M = 300$. Based on those runs, we found RE = 0.0266.

Note that, in this case, the estimator of $|\mathcal{X}^*|$ is very large and, thus, full enumeration is impossible. We made, however, the exact solution $|\mathcal{X}^*|$ available as well. It is $|\mathcal{X}^*| = (1,346,963)^4 = 3.297\text{E}+24$ and is obtained using a specially designed procedure (see Remark 5.6, below) for SAT instances generation. In particular, the instance matrix 300 × 1080 was generated from the previous one 75 × 270, for which $|\mathcal{X}^*| = 1,346,963$.

Table 5.15 Performance of SE Algorithm 5.3 for the 3-SAT 300 × 1080 model with $N_t^{(e)} = 300$, $M = 300$ and $r = 1$

Run	Iterations	$\widehat{\lvert\mathcal{X}^*\rvert}$	CPU
1	300	3.30E+24	2010.6
2	300	3.46E+24	2271.8
3	300	3.40E+24	2036.8
4	300	3.42E+24	2275.8
5	300	3.39E+24	2022.4
6	300	3.35E+24	2267.8
7	300	3.34E+24	2019.6
8	300	3.34E+24	2255.4
9	300	3.32E+24	2031.7
10	300	3.33E+24	2149.6
Average	300	3.36E+24	2134.1

Remark 5.6 ***Generating an Instance with an Available Solution***

We shall show that, given a small SAT instance with a known solution $|\mathcal{X}^*|$, we can generate an associated instance of a large size and still obtain its exact solution. Suppose without loss of generality that we have k instances (of relatively small sizes) with known $|\mathcal{X}_i^*|$, $i = 1, \ldots, k$. Denote those instances by $I_1, \ldots, I_k$, their dimensionality by $(n_1, m_1), \ldots, (n_k, m_k)$ and their corresponding solutions by $|\mathcal{X}_1^*|, \ldots, |\mathcal{X}_k^*|$. We will show how to construct a new SAT instance that will have a size $(\sum_{i=1}^k n_i, \sum_{i=1}^k m_i)$ and its exact solution will be equal to $\prod_{i=1}^k |\mathcal{X}_i^*|$. The idea is very simple. Indeed, denote the variables of I_1 by $x_1, \ldots, x_{n_1}$. Now, take the second instance and rename its variables from $x_1, \ldots, x_{n_2}$ to $x_{n_1+1}, \ldots, x_{n_1+n_2}$; that is to each variable index of I_2 we add n_1 new variables. Continue in the same manner with the rest of the instances. It should be clear that we have now an instance of size $(\sum_{i=1}^k n_i, \sum_{i=1}^k m_i)$.

Let us look next at some particular solution

$$X_1, \ldots, X_{n_1}, X_{n_1+1}, \ldots, X_{n_1+n_2}, \ldots, X_{\sum_{i=1}^k n_i}$$

of this instance. This solution consists of independent components of sizes $n_1, \ldots, n_k$, and it is straightforward to see that the total number of those solutions is $\prod_{i=1}^k |\mathcal{X}_i^*|$. It follows, therefore, that one can easily construct a large SAT instance from a set of small ones and still have an exact solution for it. Note that the above 300 × 1080 instance with exactly $(1,346,963)^4$ solutions was obtained from the four identical instances of size 75 × 270, each with exactly 1,346,963 solutions.

We also performed experiments with different values of r. Table 5.16 summarizes the results. In particular, it presents the relative error for $r_1 = 1$, $r_3 = 3$,

Table 5.16 The relative errors as functions of r

Instance	$r = 1$	$r = 2$	$r = 3$
20×80, $N_t^{(e)} = 3$, $M = 100$	2.284E-02	1.945E-02	2.146E-02
75×325, $N_t^{(e)} = 20$, $M = 100$	5.057E-02	4.587E-02	5.614E-02
75×270, $N_t^{(e)} = 100$, $M = 100$	4.449E-02	4.745E-02	4.056E-02

and $r_5 = 5$ with SE run for a predefined time period for each instance. We can see that changing r does not affect the relative error.

5.5.5 Comparison of SE with Splitting and SampleSearch

Here we compare SE with splitting and SampleSearch for several 3-SAT instances. Before doing so we require the following remarks.

Remark 5.7 *SE versus Splitting*

1. In the splitting method of Chapter 4, the rare event ℓ is presented as

$$\ell = \mathbb{E}\left[I_{\{\sum_{j=1}^{m} C_j(\boldsymbol{X})=m\}}\right], \tag{5.19}$$

where $\boldsymbol{X}$ has a uniform distribution on a finite n-dimensional set $\mathcal{X}_0$, and m is the number of clauses C_j, $j = 1, \ldots, m$. To estimate $|\mathcal{X}^*|$ the splitting algorithm generates an adaptive sequence of pairs

$$\{(m_0, g^*(\boldsymbol{x}, m_0)), \quad (m_1, g^*(\boldsymbol{x}, m_1)), \quad (m_2, g^*(\boldsymbol{x}, m_2)), \ldots, (m_T, g^*(\boldsymbol{x}, m_T))\}, \tag{5.20}$$

where $g^*(\boldsymbol{x}, m_t)$, $t = 0, 1, \ldots, T$ is uniformly distributed on the set $\mathcal{X}_t$ and such that $\mathcal{X}_0 \supset \mathcal{X}_1 \supset \cdots \supset \mathcal{X}_T = \mathcal{X}^*$.

2. In contrast to splitting, which samples from a sequence of n-dimensional pdf's $g^*(\boldsymbol{x}, m_t)$ (see (5.20)), sampling in SE Algorithm 5.3 is minimal; it resorts to sampling only n times. In particular, SE draws $N_t^{(e)}$ balls (without replacing) from an urn containing $N_t \geq N_t^{(e)}$, $t = 1, \ldots, n$ ones.
3. Splitting relies on the time-consuming MCMC method and in particular on the Gibbs sampler, whereas SE dispenses with them and is thus substantially simpler and faster. For more details on splitting, see Chapter 4.
4. The limitation of SE relative to splitting is that the former is suitable for counting, where only fast (polynomial) decision-making oracles are available, whereas splitting dispenses with them. In addition, it is not suitable for optimization and rare events, as splitting is.

Table 5.17 Comparison of the efficiencies of SE, SampleSearch and standard splitting

Instance	Time	SaS	SaS RE	SE	SE RE	Split	Split RE
20 × 80	1 sec	14.881	7.95E-03	15.0158	5.51E-03	14.97	3.96E-02
75 × 325	137 sec	2212	2.04E-02	2248.8	9.31E-03	2264.3	6.55E-02
75 × 270	122 sec	1.32E+06	2.00E-02	1.34E+06	1.49E-02	1.37E+06	3.68E-02
300 × 1080	1600 sec	1.69E+23	9.49E-01	3.32E+24	3.17E-02	3.27E+24	2.39E-01

Remark 5.8 *SE versus SampleSearch*

The main difference between `SampleSearch` [56, 57] and SE is that the former approximates an entire #P-complete counting problem like SAT by incorporating IS in the oracle. As a result, it is only asymptotically unbiased. Wei and Selman's [126] `ApproxCount` is similar to `SampleSearch` in that it uses an MCMC sampler instead of IS. In contrast to both of the above, SE reduces the difficult counting problem to a set of simple ones, applying the oracle each time directly (via the **Full Enumeration** step) to the elite trajectories of size $N_t^{(e)}$. Note also that, unlike both of the above, there is no randomness involved in SE as far as the oracle is concerned in the sense that, once the elite trajectories are given, the oracle generates (via **Full Enumeration**) a deterministic sequence of new trajectories of size N_t. As a result, SE is unbiased for any $N_t^{(e)} \geq 1$ and is typically more accurate than `SampleSearch`. Our numerical results below confirm this.

Table 5.17 presents a comparison of the efficiencies of SE (at $r = 1$) with those of `SampleSearch` and splitting (SaS and Split columns, respectively) for several SAT instances. We ran all three methods for the same amount of time.

In terms of speed (which equals $(\text{RE})^2$), SE is faster than `SampleSearch` by about 2–10 times and standard splitting in [104] by about 20–50 times. Similar comparison results were obtained for other models, including perfect matching. Our explanation is that SE is an SIS method, whereas `SampleSearch` and splitting are not, in the sense that SE samples sequentially with respect to coordinates $x_1, \ldots, x_n$, whereas the other two sample (random vectors $\mathbf{X}$ from the IS pdf $g(\boldsymbol{x})$) in the entire n-dimensional space.

Appendix A

Additional Topics

A.1 COMBINATORIAL PROBLEMS

Combinatorics is a branch of mathematics concerning the study of finite or countable discrete structures and, in particular, how the discrete structures can be combined or arranged. Aspects of combinatorics include counting the structures of a given kind and size, deciding when certain criteria can be met, and combinatorial optimization.

Combinatorial problems arise in many areas of pure mathematics, notably in algebra, probability theory, topology, and geometry, and combinatorics also has many applications in optimization, computer science, ergodic theory, and statistical physics. In the later twentieth century, however, powerful and general theoretical methods were developed, making combinatorics into an independent branch of mathematics in its own right. Combinatorics is used frequently in computer science to obtain formulas and estimates in the analysis of algorithms.

An important part of the study is to analyze the complexity of the proposed algorithms. In combinatorics, we are interested in particular arrangements or combinations of the objects. For instance, neighboring countries are colored differently when drawn on a map. We might be interested in the minimal number of colors needed or in how many different ways we can color the map when given a number of colors to use.

Knuth [68, page 1] distinguishes five issues of concern in combinatorics:

i. Existence: Is there an arrangement?

ii. Construction: If yes, can we find one quickly?

iii. Enumeration: How many different arrangements are there?

iv. Generation: Can all arrangements be visited systematically?

v. Optimization: What arrangement is optimal, given an objective function on the set of arrangements?

Fast Sequential Monte Carlo Methods for Counting and Optimization, First Edition.
Reuven Y. Rubinstein, Ad Ridder, and Radislav Vaisman.

In this book we focus on issues (i), (iii), and (v). Issue (i) is also called the decision problem; issue (iii) is also known as the counting problem. In particular, we consider combinatorial problems that can be modeled by an optimization program and, in fact, most often by an integer linear program. For that purpose, we denote by $\mathcal{X}^*$ the set of feasible solutions of the problem, and we assume that it is a subset of an n-dimensional integer vector space and that is given by the following linear integer constraints (cf. (1.1)):

$$\mathcal{X}^* = \begin{cases} \sum_{j=1}^n a_{ij} x_j = b_i, & \text{for all } i = 1, \ldots, m_1 \\ \sum_{j=1}^n a_{ij} x_j \geq b_i, & \text{for all } i = m_1 + 1, \ldots, m_1 + m_2 = m \\ \boldsymbol{x} = (x_1, \ldots, x_n) \in \mathbb{Z}^n \end{cases} \tag{A.1}$$

where,

1. $\boldsymbol{A} = (a_{ij})$ is a $m \times n$ matrix of constraint coefficients, and $\boldsymbol{b} = (b_i)$ is a m-vector of the right-hand side values. Most often we assume that the variables x_j are non-negative integers.
2. The combinatorial problems associated with (A.1) are
   - *Decision making: is $\mathcal{X}^*$ nonempty?*
   - *Counting: calculate $|\mathcal{X}^*|$.*
   - *Optimization: solve* $\max_{\boldsymbol{x} \in \mathcal{X}^*} S(\boldsymbol{x})$ *for a given objective or performance function* $S : \mathcal{X}^* \to \mathbb{R}$.

In this book, we describe in detail various problems, algorithms, and mathematical aspects that are related to the issues (i)–(iii) and associated with the problem (A.1). The following two sections deal with a sample of well-known counting and optimization problems, respectively.

A.1.1 Counting

In this section, we present details of the satisfiability problem, independent set problem, vertex coloring problem, Hamiltonian cycle problem, the knapsack problem, and the permanent problem.

A.1.1.1 Satisfiability Problem

The most common satisfiability problem (SAT) comprises the following two components:

- A set of n boolean variables $\{x_1, \ldots, x_n\}$, representing statements that can either be TRUE (= 1) or FALSE (= 0). The negation (the logical NOT) of a variable x_j is denoted by $\overline{x}_j$. A variable or its negation is called a literal.
- A set of m distinct clauses $\{C_1, C_2, \ldots, C_m\}$ of the form $C_i = l_{i_1} \vee l_{i_2} \vee \cdots \vee l_{i_k}$, where the l's are the literals of the x_j variables, and the $\vee$ denotes the logical OR operator.

The binary vector $\boldsymbol{x} = (x_1, \ldots, x_n)$ is called a truth assignment, or simply an assignment. Thus, $x_j = 1$ assigns truth to x_j and $x_j = 0$ assigns truth to $\overline{x}_j$, for each $i = 1, \ldots, n$. The simplest SAT problem can now be formulated as follows: find a truth assignment $\boldsymbol{x}$ such that all clauses are true.

Denoting the logical AND operator by $\wedge$, we can represent the above SAT problem via a single formula as

$$F(\boldsymbol{x}) = \bigwedge_{1 \le i \le m} \left(\bigvee_{1 \le k \le |C_i|} l_{i_k} \right)$$

where $|C_i|$ is the number of literals in clause C_i and l_{i_k} are the literals of the x_j variables in clause C_i. The SAT formula is then said to be in conjunctive normal form (CNF). When all clauses have exactly the same number k of literals per clause, we say that we deal with the k-SAT problem.

The problem of deciding whether there exists a valid assignment, and, indeed, providing such a vector is called the SAT-assignment problem and is NP-complete [92]. We are concerned with the harder problem to count all valid assignments, which is $\#P$-complete.

SAT problems can be converted into the framework of a solution set given by linear constraints. Define the $m \times n$ matrix $\boldsymbol{A}$ with entries $a_{ij} \in \{-1, 0, 1\}$ by

$$a_{ij} = \begin{cases} -1 & \text{if } \bar{x}_j \in C_i \\ 0 & \text{if } x_j \notin C_i \text{ and } \bar{x}_j \notin C_i \\ 1 & \text{if } x_j \in C_i . \end{cases}$$

Furthermore, let $\boldsymbol{b}$ be the m-vector with entries $b_i = 1 - |\{j : a_{ij} = -1\}|$. Then it is easy to see that, for any configuration $\boldsymbol{x} \in \{0, 1\}^n$,

$$\boldsymbol{x} \in \mathcal{X}^* \Leftrightarrow F(\boldsymbol{x}) = 1 \Leftrightarrow \boldsymbol{A}\boldsymbol{x} \ge \boldsymbol{b}.$$

Thus, the SAT problem presents a particular case of the integer linear program with constraints given in (A.1).

A.1.1.2 Independent Sets

Consider a graph $G = G(V, E)$ with n nodes and m edges. Our goal is to count the number of independent node sets of this graph. A node set is called independent if no two nodes are connected by an edge, that is, no two nodes are adjacent; see Figure A.1 for an illustration of this concept.

The independent set problem can be converted into the family of integer programming problems (A.1) by considering the transpose of the incidence matrix $\boldsymbol{A}$. The nodes are labeled $v_1, \ldots, v_n$; the edges are labeled $e_1, \ldots, e_m$. Then $\boldsymbol{A}$ is a $m \times n$ matrices of zeros and ones with $a_{ij} = 1$ iff the edge e_i and node v_j are incident. The decision vector $\boldsymbol{x} \in \{0, 1\}^n$ denotes which nodes belong to the

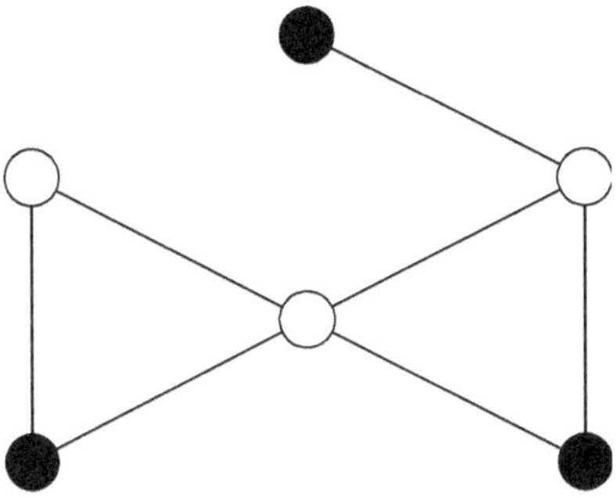

Figure A.1 The black nodes form an independent set inasmuch as they are not adjacent to each other.

independent set. Finally, let $\boldsymbol{b}$ be the m-column vector with entries $b_i = 1$ for all i. Then it is easy to see that for any configuration $\boldsymbol{x} \in \{0, 1\}^n$

$$\boldsymbol{x} \in \mathcal{X}^* \Leftrightarrow \boldsymbol{A}\boldsymbol{x} \leq \boldsymbol{b}.$$

Having formulated the problem set $\mathcal{X}^*$ as the feasible set of an integer linear program with bounded variables, we can apply the procedure mentioned in Section 4.1 for constructing a sequence of decreasing sets (4.1) in order to start the splitting method.

An alternative construction of the sequence of decreasing sets (4.1) is the following. Let $E_t = \{e_1, \ldots, e_t\}$ be the set of the first t edges and define the associated subgraph $G_t = G(V, E_t)$, $t = 1, \ldots, m$. Note that $G_m = G$, and that G_{t+1} is obtained from G_t by adding the edge e_{t+1}, which is not in G_t. Define $\mathcal{X}_t$ to be the set of the independent sets of G_t, then clearly $\mathcal{X}_0 \supset \mathcal{X}_1 \supset \cdots \supset \mathcal{X}_m = \mathcal{X}^*$. Here $|\mathcal{X}_0| = 2^n$, because G_0 has no edges and thus every subset of V is an independent set, including the empty set.

A.1.1.3 Vertex Coloring

Given a graph $G = G(V, E)$ with n nodes $v_1, \ldots, v_n$, and m edges $e_1, \ldots, e_m$, color the nodes of V with at most a q colors given from a set of colors $C = \{c_1, \ldots, c_q\}$, such that for each edge $v_i v_j \in E$, nodes v_i and v_j have different colors. We shall show how to cast the node coloring problem into the framework (A.1). Define the decision variables $x_{ik} \in \{0, 1\}$ for $i = 1, \ldots, n$ and $k = 1, \ldots, q$ by

$$x_{ik} = \begin{cases} 1 & \text{if node } v_i \text{ is colored with color } \mathrm{c}_k \\ 0 & \text{otherwise.} \end{cases}$$

Then, $\boldsymbol{x} \in \mathcal{X}^*$ iff

$$\sum_{k=1}^{c} x_{ik} = 1, \quad \text{for all } i = 1, \ldots, n$$

$$x_{ik} + x_{jk} \leq 1, \quad \text{for all } k = 1, \ldots, q, \text{ and all edges } v_i v_j \in E,\ i, j = 1, \ldots, n. \tag{A.2}$$

An alternative to (A.2) framework can be obtained by arguing as follows. Define decision variables $x_i \in \{0, \dots, q\}$ for $i = 1, \dots, n$ by

$$x_i = k \quad \text{if node } v_i \text{ is colored with color } c_k.$$

Then, $\boldsymbol{x} \in \mathcal{X}^*$ iff

$$\begin{aligned} & x_i \neq x_j, \quad \text{for all edges } v_i v_j \in E,\ i, j = 1, \dots, n \\ & x_i \in \{1, \dots, q\}, \quad \text{for all } i = 1, \dots, n. \end{aligned} \tag{A.3}$$

The procedure for node coloring via the randomized algorithm is the same as for the independent sets problem. Again, let E_t be the set of the first t edges and let $G_t = G(V, E_t)$ be the associated subgraph, $t = 1, \dots, m$. Note that $G_m = G$, and that G_{t+1} is obtained from G_t by adding the edge e_{t+1}. Define $\mathcal{X}_t$ to be the set of the node colorings of G_t, then clearly $\mathcal{X}_0 \supset \mathcal{X}_1 \supset \cdots \supset \mathcal{X}_m = \mathcal{X}^*$. Here $|\mathcal{X}_0| = q^n$, because G_0 has no edges and thus we can color any node with any color.

A.1.1.4 Hamiltonian Cycles

Given a graph $G = G(V, E)$ with n nodes and m edges, find all Hamiltonian cycles. Figure A.2 presents a graph with eight nodes and several Hamiltonian cycles, one of which is marked in bold lines.

A.1.1.5 Permanent

Let $\boldsymbol{A}$ be a $n \times n$ matrix consisting of zeros and ones only. The permanent of $\boldsymbol{A}$ is defined by

$$\text{per}(\boldsymbol{A}) = \sum_{\sigma \in \Sigma_n} \prod_{i=1}^{n} a_{i\sigma_i}, \tag{A.4}$$

where Σ_n is the set of all permutations $\sigma = (\sigma_1, \dots, \sigma_n)$ of $(1, \dots, n)$.

It is well known that the permanent of $\boldsymbol{A}$ equals the number of perfect matchings in a balanced bipartite graph $G = G((V_1, V_2), E)$ with matrix $\boldsymbol{A}$ as its biadjacency matrix [90]: two disjoint sets of nodes $V_1 = (v_{11}, \dots, v_{1n})$ and $V_2 =$

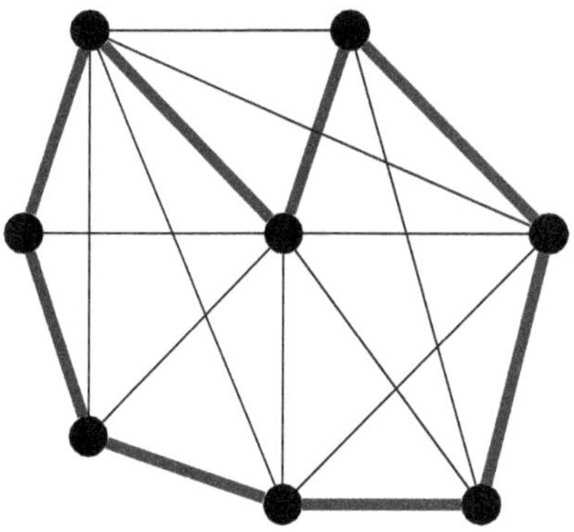

Figure A.2 A Hamiltonian graph. The bold edges form a Hamiltonian cycle.

$(v_{21}, \ldots, v_{2n})$, where $v_{1i}v_{2j} \in E$ iff $a_{ij} = 1$. Recall that a matching is a collection of edges $M \subseteq E$ such that each node occurs at most once in M. A perfect matching is a matching of size n.

To cast the bipartite perfect matching problem in an integer linear program, we define the decision variables $(x_{ij})_{i,j=1,\ldots,n}$ by

$$x_{ij} = \begin{cases} 1 & \text{if edge } (v_{1i}, v_{2j}) \in M \\ 0 & \text{otherwise.} \end{cases}$$

Let $\mathcal{X}^*$ be the set of all perfect matchings of the bipartite graph associated with the given binary matrix $\boldsymbol{A}$. Then $\boldsymbol{x} \in \mathcal{X}^*$ iff

$$\begin{aligned} &\sum_{j=1}^{n} x_{ij} = 1, \quad \text{for all } \ i = 1, \ldots, n \\ &\sum_{i=1}^{n} x_{ij} = 1, \quad \text{for all } \ j = 1, \ldots, n \\ &x_{ij} \in \{0, 1\}, \quad \text{for all } \ i, j = 1, \ldots, n. \end{aligned} \tag{A.5}$$

As an example, let $\boldsymbol{A}$ be the 3×3 matrix

$$\boldsymbol{A} = \begin{pmatrix} 1 & 1 & 1 \\ 1 & 1 & 0 \\ 0 & 1 & 1 \end{pmatrix}. \tag{A.6}$$

The corresponding bipartite graph is given in Figure A.3. The graph has three perfect matchings, one of which is displayed in the figure.

A.1.1.6 Knapsack Problem

Given items of sizes $a_1, \ldots, a_n > 0$ and a positive integer $b \geq \min_j a_j$, find the numbers of vectors $\boldsymbol{x} = (x_1, \ldots, x_n) \in \{0, 1\}^n$, such that

$$\sum_{j=1}^{n} a_j \, x_j \leq b.$$

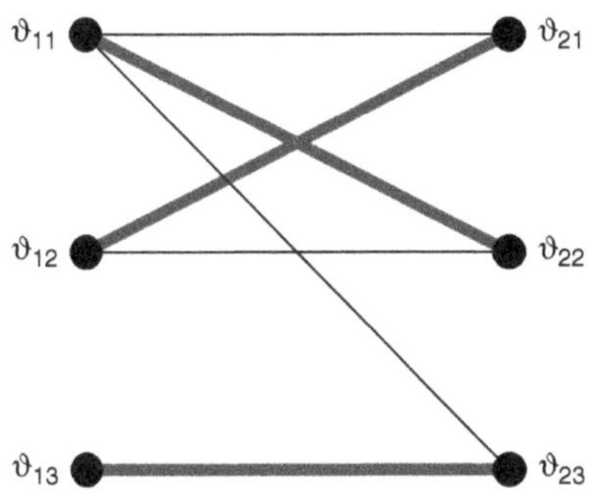

Figure A.3 A bipartite graph. The bold edges form a perfect matching.

The integer b represents the size of the knapsack, and x_j indicates whether or not item j is placed in it. Let $\mathcal{X}^*$ denote the set of all feasible solutions, that is, all different combinations of items that can be placed in the knapsack without exceeding its capacity. The goal is to determine $|\mathcal{X}^*|$.

A.1.2 Combinatorial Optimization

In this section, we present details of the max-cut problem, the traveling salesman problem, Internet security problem, set covering/partitioning/packing problem, knapsack problems, the maximum satisfiability problem, the weighted matching problem, and graph coloring problems.

A.1.2.1 Max-Cut Problem

The maximal-cut (max-cut) problem in a graph can be formulated as follows. Given a graph $G = G(V, E)$ with a set of nodes $V = \{v_1, \ldots, v_n\}$, a set of edges E between the nodes, and a weight function on the edges, $c : E \to \mathbb{R}$. The problem is to partition the nodes into two subsets V_1 and V_2 such that the sum of the weights c_{ij} of the edges going from one subset to the other is maximized:

$$\max\left\{ \sum_{\substack{v_i \in V_1 \\ v_j \in V_2}} c_{ij} \; : \; V_1 \cup V_2 = V, V_1 \cap V_2 = \emptyset \right\}.$$

A cut can be conveniently represented via its corresponding cut vector $\boldsymbol{x} = (x_1, \ldots, x_n)$, where $x_i = 1$ if node v_i belongs to same partition as v_1, and $x_i = 0$ else. For each cut vector $\boldsymbol{x}$, let $\{V_1(\boldsymbol{x}), V_2(\boldsymbol{x})\}$ be the partition of V induced by $\boldsymbol{x}$, such that $V_1(\boldsymbol{x})$ contains the set of nodes $\{v_i : x_i = 1\}$. Unless stated otherwise, we let $v_1 \in V_1$ (thus, $x_1 = 1$).

Let $\mathcal{X}$ be the set of all cut vectors $\boldsymbol{x} = (1, x_2, \ldots, x_n)$ and let $S(\boldsymbol{x})$ be the corresponding cost of the cut. Then,

$$S(\boldsymbol{x}) = \sum_{\substack{v_i \in V_1(\boldsymbol{x}) \\ v_j \in V_2(\boldsymbol{x})}} c_{ij}. \tag{A.7}$$

When we do not specify, we assume that the graph is undirected. However, when we deal with a directed graph, the cost of a cut $\{V_1, V_2\}$ includes both the cost of the edges from V_1 to V_2 and from V_2 to V_1. In this case, the cost corresponding to a cut vector $\boldsymbol{x}$ is, therefore,

$$S(\boldsymbol{x}) = \sum_{\substack{v_i \in V_1(\boldsymbol{x}) \\ v_j \in V_2(\boldsymbol{x})}} (c_{ij} + c_{ji}). \tag{A.8}$$

A.1.2.2 Traveling Salesman Problem

The traveling salesman problem (TSP) can be formulated as follows. Consider a weighted graph G with n nodes, labeled $1, 2, \ldots, n$. The nodes represent cities, and the edges represent the roads between the cities. Each edge from i to j has weight or cost c_{ij}, representing the length of the road. The problem is to find the shortest tour that visits all the cities exactly once and such that the starting city is also the terminating one.

Let $\mathcal{X}$ be the set of all possible tours and let $S(\boldsymbol{x})$ be the total length of tour $\boldsymbol{x} \in \mathcal{X}$. We can represent each tour via a permutation $\boldsymbol{x} = (x_1, \ldots, x_n)$ with $x_1 = 1$. Mathematically, the TSP reads as

$$\min_{\boldsymbol{x}\in\mathcal{X}} S(\boldsymbol{x}) = \min_{\boldsymbol{x}\in\mathcal{X}} \left\{ \sum_{i=1}^{n-1} c_{x_i,x_{i+1}} + c_{x_n,1} \right\}. \tag{A.9}$$

A.1.2.3 RSA or Internet Security Problem

The RSA, also called the Internet security problem, reads as follows: find two primes, provided we are given a large integer m known to be a product of these primes. Clearly, we can write the given number m in the binary system with $n + 1$-bit integers as

$$m = \alpha_0 + 2\alpha_1 + \cdots + 2^n\alpha_n, \quad \text{where } \alpha_j \in \{0, 1\},\ j = 1, \ldots, n.$$

Mathematically, the problem reads as follows: find binary $x_i, y_j, i, j = 0, 1, \ldots, n$ such that

$$\begin{aligned} &\sum_{k=0}^{2n} 2^k \sum_{i=0}^{k} x_i y_{k-i} = \sum_{i,j=0}^{n} 2^{i+j} x_i y_j = \left(\sum_{i=0}^{n} 2^i x_i\right)\left(\sum_{j=0}^{n} 2^j y_j\right) = m \\ &x_i \in \{0, 1\}, \quad y_i \in \{0, 1\}, \quad \text{for all } i, j = 0, 1, \ldots, n. \end{aligned} \tag{A.10}$$

Remarks:

- The problem (A.10) can be formulated for any basis, say for a decimal rather than a binary one.
- To have a unique solution, we can impose in addition the constraint $\sum_{i=0}^{n} 2^i x_i > \sum_{j=0}^{n} 2^j y_j$.
- The RSA problem can also be formulated as:
  Find binary $x_i, y_j, i, j = 0, 1, \ldots, n$ such that

$$\begin{aligned} &\sum_{i,j=0}^{n} 2^{i+j} x_i y_j = m \\ &2x_i y_j \le x_i + y_j, \quad \text{for all } i, j = 0, 1, \ldots, n \\ &x_i y_j \ge x_i + y_j - 1, \quad \text{for all } i, j = 0, 1, \ldots, n \\ &x_i \in \{0, 1\}, \quad y_i \in \{0, 1\}, \quad \text{for all } i, j = 0, 1, \ldots, n. \end{aligned} \tag{A.11}$$

EXAMPLE A.1

Let $m = 77$. In this case, the primes are $b_1 = 7$ and $b_2 = 11$. We can write them as

$$b_1 = \sum_i 2^i x_i = 1 \cdot 2^0 + 1 \cdot 2^1 + 1 \cdot 2^2 + 0 \cdot 2^3 = 7$$

and

$$b_2 = \sum_j 2^j y_j = 1 \cdot 2^0 + 1 \cdot 2^1 + 0 \cdot 2^2 + 1 \cdot 2^3 = 11,$$

respectively. In this case $n = 3$ and the program (A.10) reduces to:

Find binary x_i, y_j, z_{ij}, $i, j = 0, 1, \ldots, 3$ such that

$$\begin{aligned} &\sum_{i,j} 2^{i+j} z_{ij} = 77 \\ &2z_{ij} \leq x_i + y_j, \quad \text{for all } i, j = 0, 1, \ldots, 3 \\ &z_{ij} \geq x_i + y_j - 1, \quad \text{for all } i, j = 0, 1, \ldots, 3 \\ &x_i \in \{0, 1\},\ y_i \in \{0, 1\},\ z_{ij} \in \{0, 1\}, \quad \text{for all } i, j = 0, 1, \ldots, 3. \end{aligned} \tag{A.12}$$

The program should deliver the unique solution

$$\boldsymbol{x} = \{1,\ 1,\ 1,\ 0\},\ \boldsymbol{y} = \{1,\ 1,\ 0,\ 1\}.$$

Note that we should consider $\boldsymbol{x} = \{1,\ 1,\ 0,\ 1\}$ and $\boldsymbol{y} = \{1,\ 1,\ 1,\ 0\}$ as the same solution as the previous one. □

A.1.2.4 Set Covering, Set Packing, and Set Partitioning

Consider a finite set $\{M = 1, 2, \ldots, m\}$ and let M_j, $j \in N$ be a collection of subsets of the set M where $N = \{1, 2, \ldots, n\}$. A subset $F \subseteq N$ is called a cover of M if $\cup_{j \in F} M_j = M$, and is called a packing of M if $M_j \cap M_k = \emptyset$ for all $j, k \in F$ and $j \neq k$. If $F \subseteq N$ is both a cover and packing, it is called a partitioning.

Suppose c_j is the cost associated with M_j. Then the set covering problem is to find a minimum cost cover. If c_j is the value or weight of M_j, then the set packing problem is to find a maximum weight or value packing. Similarly, the set partitioning problem is to find a partitioning with minimum cost. These problems can be formulated as zero-one linear integer programs as shown below. For all $i \in M$ and $j \in N$, let

$$a_{ij} = \begin{cases} 1 & \text{if } i \in M_j \\ 0 & \text{otherwise} \end{cases}$$

and

$$x_j = \begin{cases} 1 & \text{if } j \in F \\ 0 & \text{otherwise.} \end{cases}$$

Then, the set covering, set packing, and set partitioning formulations are given by

$$\min \sum_{j=1}^{n} c_j x_j$$
$$\text{s.t.} \sum_{j=1}^{n} a_{ij} x_j \geq 1, \quad \text{for all } i = 1, 2, \dots, m$$
$$x_j \in \{0, 1\}, \quad \text{for all } j = 1, 2, \dots, n,$$

$$\max \sum_{j=1}^{n} c_j x_j$$
$$\text{s.t.} \sum_{j=1}^{n} a_{ij} x_j \leq 1, \quad \text{for all } i = 1, 2, \dots, m$$
$$x_j \in \{0, 1\}, \quad \text{for all } j = 1, 2, \dots, n,$$

and

$$\max \sum_{j=1}^{n} c_j x_j$$
$$\text{s.t.} \sum_{j=1}^{n} a_{ij} x_j = 1, \quad \text{for all } i = 1, 2, \dots, m$$
$$x_j \in \{0, 1\}, \quad \text{for all } j = 1, 2, \dots, n,$$

respectively.

A.1.2.5 Maximum Satisfiability Problem

The satisfiability problem has been introduced in Section A.1.1. When there is no feasible solution, one might be interested in finding an assignment of truth values to the variables that satisfies the maximum number of clauses, called the MAX-SAT problem.

If one associates a weight w_i to each clause C_i, one obtains the weighted MAX-SAT problem, denoted by MAX W-SAT. The assignment of truth values to the n variables that maximizes the sum of the weights of the satisfied clauses is determined. Of course, MAX-SAT is contained in MAX W-SAT (all weights are equal to one). When all clauses have exactly the same number k of literals per clause, we say that we deal with the MAX-k-SAT problem, or MAX W-k-SAT in the weighted case.

The MAX W-SAT problem has a natural integer linear programming formulation (ILP). Recall the boolean decision variables $x_j \in \{0, 1\}$, meaning TRUE ($x_j = 1$) and FALSE ($x_j = 0$). Let the boolean variable $z_i = 1$ if clause C_i is satisfied, $z_i = 0$ otherwise. The integer linear program is

$$\max \sum_{i=1}^{m} w_i z_i$$

$$
\text{s.t.} \sum_{j \in U_i^+} x_j + \sum_{j \in U_i^-} (1 - x_j) \geq z_i, \quad \text{for all } i = 1, \ldots, m
$$
$$
x_j \in \{0, 1\}, \quad \text{for all } j = 1, \ldots, n
$$
$$
z_i \in \{0, 1\}, \quad \text{for all } i = 1, \ldots, m,
$$

where U_i^+ and U_i^- denote the set of indices of variables that appear un-negated and negated in clause C_i, respectively.

Because the sum of the $z_i w_i$ is maximized and because each z_i appears as the right-hand side of one constraint only, z_i will be equal to one if and only if clause C_i is satisfied.

A.1.2.6 Weighted Matching

Let $G = (V, E)$ be a graph with n nodes (n even), and let $c : E \to \mathbb{R}$ be a weight function on the edges, then the (general weighted) matching problem is formulated as

$$
\begin{aligned}
\min \quad & \sum_{i,j;v_i v_j \in E = 1} c_{ij} x_{ij} \\
\text{s.t.} \quad & \sum_{j=1}^{n} x_{ij} = 1, \quad \text{for all } i = 1, \ldots, n \\
& \sum_{i=1}^{n} x_{ij} = 1, \quad \text{for all } i = 1, \ldots, n \\
& x_{ij} \in \{0, 1\}, \quad \text{for all } i, j = 1, \ldots, n.
\end{aligned} \tag{A.13}
$$

A.1.2.7 Knapsack Problems

All knapsack problems consider a set of items with associated profit p_j and weight w_j. A subset of the items is to be chosen such that the weight sum does not exceed the capacity c of the knapsack, and such that the largest possible profit sum is obtained. We will assume that all coefficients p_j, w_j, c are positive integers, although weaker assumptions may sometimes be handled in the individual problems.

The 0-1 knapsack problem is the problem of choosing some of the n items such that the corresponding profit sum is maximized without the weight sum exceeding the capacity c. Thus, it may be formulated as the following maximization problem:

$$
\begin{aligned}
\max \quad & \sum_{j=1}^{n} p_j x_j \\
\text{s.t.} \quad & \sum_{j=1}^{n} w_j x_j \leq c \\
& x_j \in \{0, 1\}, \quad \text{for all } j = 1, \ldots, n,
\end{aligned} \tag{A.14}
$$

where x_j is a binary variable having value 1 if item j should be included in the knapsack, and 0 otherwise. If we have a bounded amount m_j of each item type j, then the bounded knapsack problem appears:

$$\begin{aligned} \max & \sum_{j=1}^{n} p_j x_j \\ \text{s.t.} & \sum_{j=1}^{n} w_j x_j \le c \\ & x_j \in \{0, 1, \dots, m_j\}, \quad \text{for all } j = 1, \dots, n. \end{aligned} \tag{A.15}$$

Here x_j gives the amount of each item type that should be included in the knapsack in order to obtain the largest objective value. The unbounded knapsack problem is a special case of the bounded knapsack problem, since an unlimited amount of each item type is available:

$$\begin{aligned} \max & \sum_{j=1}^{n} p_j x_j \\ \text{s.t.} & \sum_{j=1}^{n} w_j x_j \le c \\ & x_j \ge 0 \text{ integer}, \quad \text{for all } j = 1, \dots, n. \end{aligned} \tag{A.16}$$

Actually, any variable x_j of an unbounded knapsack problem will be bounded by the capacity c, as the weight of each item is at least one. But, generally, there is no benefit by transforming an unbounded knapsack problem into its bounded version.

The most general form of a knapsack problem is the multiconstrained knapsack problem, which is basically a general integer programming problem, where all coefficients p_j, w_{ij}, and c_i are non-negative integers. Thus, it may be formulated as

$$\begin{aligned} \max & \sum_{j=1}^{n} p_j x_j \\ \text{s.t.} & \sum_{j=1}^{n} w_{ij} x_j \le c_i, \quad \text{for all } i = 1, \dots, m \\ & x_j \ge 0 \text{ integer}, \quad \text{for all } j = 1, \dots, n. \end{aligned} \tag{A.17}$$

The quadratic knapsack problem presented is an example of a knapsack problem with a quadratic objective function. It may be stated as

$$\max \sum_{j=1}^{n} \sum_{i=1}^{n} p_{ij} x_i x_j$$

$$\text{s.t.} \sum_{j=1}^{n} w_j x_j \leq c \tag{A.18}$$

$$x_j \in \{0, 1\}, \quad \text{for all } j = 1, \ldots, n.$$

Here, p_{ij} is the profit obtained if both items i and j are chosen, while w_j is the weight of item j. The quadratic knapsack problem is a knapsack counterpart to the quadratic assignment problem, and the problem has several applications in telecommunication and hydrological studies.

A.1.2.8 Graph Coloring

Similar to many combinatorial optimization problems, the graph coloring problem has several mathematical programming formulations. Five such formulations are presented in the following:

(F-1):

$$\min \sum_{k=1}^{n} y_k$$

$$\text{s.t.} \sum_{k=1}^{n} x_{ik} = 1, \quad \text{for all nodes } v_i \in V \tag{A.19}$$

$$x_{ik} + x_{jk} \leq 1, \quad \text{for all edges } v_i v_j \in E \tag{A.20}$$

$$y_k \geq x_{ik}, \quad \text{for all nodes } v_i \in V,\ k = 1, \ldots, n \tag{A.21}$$

$$y_k,\ x_{ik} \in \{0, 1\}, \quad \text{for all nodes } v_i \in V,\ k = 1, \ldots, n. \tag{A.22}$$

In the above model, $y_k = 1$ if color k is used. The binary variables x_{ik} are associated with node v_i: $x_{ik} = 1$ if and only if color k is assigned to node v_i. Constraints (A.19) ensure that exactly one color is assigned to each node. Constraints (A.20) prevent adjacent nodes from having the same color. Constraints (A.21) guarantee that no x_{ik} can be 1 unless color k is used. The optimal objective function value gives the chromatic number of the graph. Moreover, the sets $S_k = \{i | x_{ik} = 1\}$ for all k, comprise a partition of the nodes into (minimum number of) independent sets.

(F-2):

$$\min \quad \gamma$$

$$\text{s.t.} \quad x_i \leq \gamma \tag{A.23}$$

$$x_i - x_j - 1 \geq -n\delta_{ij}, \quad \text{for all edges } v_i v_j \in E \tag{A.24}$$

$$x_j - x_i - 1 \geq -n(1 - \delta_{ij}), \quad \text{for all edges } v_i, v_j \in E \tag{A.25}$$

$$\delta_{ij} \in \{0, 1\},\ x_i \in Z^+, \quad \text{for all nodes } v_i, v_j \in V$$

The value of x_i indicates which color is assigned to v_i, for $i = 1, \ldots, n$. Constraints (A.24) and (A.25) prevent two adjacent nodes from having the same color. This can be seen by noting that, if $x_i = x_j$, then no feasible assignment of δ_{ij} will satisfy both (A.24) and (A.25). The optimal objective function value equals the chromatic number of the graph under consideration.

The coloring problem can also be formulated as a set partitioning problem. Let $S_1, S_2, \ldots, S_t$ be all the independent sets of G. Let the rows of the 0-1 matrix $\boldsymbol{A}_S$ be the characteristic vectors of S_j, $j = 1, \ldots, t$. Define variables s_j as follows:

$$s_j = \begin{cases} 1 & \text{if } S_j \text{ is a chosen color class} \\ 0 & \text{otherwise.} \end{cases}$$

$$e_{ij} = \begin{cases} 1 & \text{if node } v_i \in S_j \\ 0 & \text{otherwise.} \end{cases}$$

(F-3):

$$\begin{aligned} &\min \sum_{j=1}^{t} s_j \\ &\text{s.t. } s\boldsymbol{A}_s = 1 \\ &\quad s_j \in \{0, 1\}, \quad \text{for all } j = 1, \ldots, t, \end{aligned} \tag{A.26}$$

where $s = (s_1, \ldots, s_t)$, and $1 = (1, \ldots, 1)$ is of dimension n.

(F-4):

$$\begin{aligned} &\min \sum_{j=1}^{t} s_j \\ &\text{s.t.} \sum_{j=1}^{t} e_{ij} s_j = 1, \quad \text{for all } i = 1, \ldots, n \\ &s_j \in \{0, 1\}, \quad \text{for all } j = 1, \ldots, t. \end{aligned} \tag{A.27}$$

The graph coloring problem can also be formulated as a special case of the quadratic assignment problem. Given an integer r and a graph $G = (V, E)$, the optimal objective function value of the following problem is zero if and only if G is r colorable.

(F-5):

$$\begin{aligned} &\min \sum_{k=1}^{r} \sum_{(i,j) \in E x_{ik} x_{jk}} \\ &\text{s.t.} \sum_{k=1}^{r} x_{jk} = 1, \quad \text{for all } j = 1, \ldots, n \\ &x_{jk} \in \{0, 1\}, \quad \text{for all } j, k. \end{aligned} \tag{A.28}$$

A.2 INFORMATION

In this section we discuss briefly Shannon's entropy and Kullback–Leibler's cross-entropy. The subsequent material is taken almost verbatim from Section 1.14 [108].

A.2.1 Shannon Entropy

One of the most celebrated measures of uncertainty in information theory is the *Shannon entropy*, or simply *entropy*. The entropy of a discrete random variable X with density f is defined as

$$\mathbb{E}\left[\log_2 \frac{1}{f(X)}\right] = -\mathbb{E}[\log_2 f(X)] = -\sum_{\mathcal{X}} f(x) \log_2 f(x).$$

Here X is interpreted as a random character from an alphabet $\mathcal{X}$, such that $X = x$ with probability $f(x)$. We will use the convention $0 \log 0 = 0$.

It can be shown that the most efficient way to transmit characters sampled from f over a binary channel is to encode them such that the number of bits required to transmit x is equal to $\log_2(1/f(x))$. It follows that $-\sum_{\mathcal{X}} f(x) \log_2 f(x)$ is the expected bit length required to send a random character $X \sim f$.

A more general approach, which includes continuous random variables, is to define the entropy of a random variable X with density f by

$$\mathcal{S}(X) = -\mathbb{E}[\log f(X)] = \begin{cases} -\sum f(x) \log f(x) & \text{discrete case,} \\ -\int f(x) \log f(x)\, dx & \text{continuous case.} \end{cases} \tag{A.29}$$

Definition (A.29) can easily be extended to random vectors $\boldsymbol{X}$ as (in the continuous case)

$$\mathcal{S}(\boldsymbol{X}) = -\mathbb{E}[\log f(\boldsymbol{X})] = -\int f(\boldsymbol{x}) \log f(\boldsymbol{x}) d\boldsymbol{x} \ . \tag{A.30}$$

Often, $\mathcal{S}(\boldsymbol{X})$ is called the joint entropy of the random variables $X_1, \ldots, X_n$ and is also written as $\mathcal{S}(X_1, \ldots, X_n)$. In the continuous case, $\mathcal{S}(\boldsymbol{X})$ is frequently referred to as the differential entropy to distinguish it from the discrete case.

EXAMPLE A.2

Let X have a $\mathsf{Ber}(p)$ distribution for some $0 \le p \le 1$. The density f of X is given by $f(1) = \mathbb{P}(X = 1) = p$ and $f(0) = \mathbb{P}(X = 0) = 1 - p$, so that the entropy of X is

$$\mathcal{S}_X(p) = -p \ \log p - (1 - p) \log (1 - p) \ .$$

The graph of the entropy as a function of p is depicted in Figure A.4. Note that the entropy is maximal for $p = 1/2$, which gives the uniform density on $\{0, 1\}$.

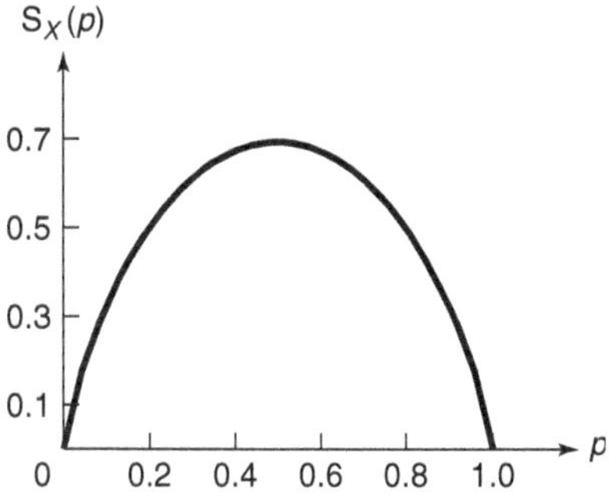

Figure A.4 The entropy for the Ber(p) distribution as a function of p.

Next, consider a sequence $X_1, \ldots, X_n$ of iid Ber(p) random variables. Let $\boldsymbol{X} = (X_1, \ldots, X_n)$. The density of $\boldsymbol{X}$, say g, is simply the product of the densities of the X_i, so that

$$\mathcal{S}(\boldsymbol{X}) = -\mathbb{E}\,[\ln\, g(\boldsymbol{X})] = -\mathbb{E}\left[\ln \prod_{i=1}^{n} f(X_i)\right] = \sum_{i=1}^{n} -\mathbb{E}\,[\ln\, f(X_i)] = n\,\mathcal{S}(X)\,.$$

The properties of $\mathcal{S}(\boldsymbol{X})$ in the continuous case are somewhat different from those in the discrete one. In particular,

1. The differential entropy can be negative, whereas the discrete entropy is always positive.
2. The discrete entropy is insensitive to invertible transformations, whereas the differential entropy is not. Specifically, if $\boldsymbol{X}$ is discrete, $\boldsymbol{Y} = g(\boldsymbol{X})$, and g is an invertible mapping, then $\mathcal{S}(\boldsymbol{X}) = \mathcal{S}(\boldsymbol{Y})$, because $f_{\boldsymbol{Y}}(\boldsymbol{y}) = f_{\boldsymbol{X}}(g^{-1}(\boldsymbol{y}))$. However, in the continuous case, we have an additional factor due to the Jacobian of the transformation.

It is not difficult to see that of any density f, the one that gives the maximum entropy is the uniform density on $\mathscr{X}$. That is,

$$\mathcal{S}(\boldsymbol{X}) \text{ is maximal } \Leftrightarrow \text{ f}(\boldsymbol{x}) = \frac{1}{|\mathscr{X}|}(\text{constant}). \tag{A.31}$$

A.2.2 Kullback–Leibler Cross-Entropy

Let g and h be two densities on $\mathscr{X}$. The Kullback–Leibler cross-entropy between g and h is defined (in the continuous case) as

$$\begin{aligned}\mathcal{D}(g,h) &= \mathbb{E}_g\left[\log \frac{g(\boldsymbol{X})}{h(\boldsymbol{X})}\right] \\ &= \int g(\boldsymbol{x}) \log g(\boldsymbol{x})\,d\boldsymbol{x} - \int g(\boldsymbol{x}) \log h(\boldsymbol{x})\,d\boldsymbol{x}\,. \end{aligned}\tag{A.32}$$

$\mathcal{D}(g, h)$ is also called the *Kullback–Leibler divergence*, the *cross-entropy*, and the *relative entropy*. If not stated otherwise, we shall call $\mathcal{D}(g, h)$ the *cross-entropy* between g and h. Notice that $\mathcal{D}(g, h)$ is not a distance between g and h in the formal sense, because, in general, $\mathcal{D}(g, h) \neq \mathcal{D}(h, g)$. Nonetheless, it is often useful to think of $\mathcal{D}(g, h)$ as a distance because

$$\mathcal{D}(g, h) \geq 0,$$

and $\mathcal{D}(g, h) = 0$ if and only if $g(x) = h(x)$. This follows from Jensen's inequality (if ϕ is a convex function, such as $-\ln$, then $\mathbb{E}[\phi(X)] \geq \phi(\mathbb{E}[X])$). Namely,

$$\mathcal{D}(g, h) = \mathbb{E}_g\left[-\log \frac{h(\boldsymbol{X})}{g(\boldsymbol{X})}\right] \geq -\log \left\{\mathbb{E}_g\left[\frac{h(\boldsymbol{X})}{g(\boldsymbol{X})}\right]\right\} = -\log 1 = 0 .$$

A.3 EFFICIENCY OF ESTIMATORS

In this book, we frequently use

$$\widehat{\ell} = \frac{1}{N}\sum_{i=1}^{N} Z_i , \tag{A.33}$$

which presents an unbiased estimator of the unknown quantity $\ell = \mathbb{E}[Z]$, where $Z_1, \ldots, Z_N$ are independent replications of some random variable Z.

By the central limit theorem, $\widehat{\ell}$ has approximately a $\mathsf{N}(\ell, N^{-1}\mathbb{V}\mathrm{ar}[Z])$ distribution for large N. We shall estimate $\mathbb{V}\mathrm{ar}[Z]$ via the *sample variance*

$$S^2 = \frac{1}{N-1}\sum_{i=1}^{N} (Z_i - \widehat{\ell})^2 .$$

By the law of large numbers, S^2 converges with probability 1 to $\mathbb{V}\mathrm{ar}[Z]$. Consequently, for $\mathbb{V}\mathrm{ar}[Z] < \infty$ and large N, the approximate $(1 - \alpha)$ confidence interval for ℓ is given by

$$\left(\widehat{\ell} - z_{1-\alpha/2}\,\frac{S}{\sqrt{N}},\ \widehat{\ell} + z_{1-\alpha/2}\,\frac{S}{\sqrt{N}}\right),$$

where $z_{1-\alpha/2}$ is the $(1 - \alpha/2)$ quantile of the standard normal distribution. For example, for $\alpha = 0.05$ we have $z_{1-\alpha/2} = z_{0.975} = 1.96$. The quantity

$$\frac{S/\sqrt{N}}{\widehat{\ell}}$$

is often used in the simulation literature as an accuracy measure for the estimator $\widehat{\ell}$. For large N it converges to the *relative error* of $\widehat{\ell}$, defined as

$$\mathrm{RE} = \frac{\sqrt{\mathbb{V}\mathrm{ar}[\widehat{\ell}]}}{\mathbb{E}[\widehat{\ell}]} = \frac{\sqrt{\mathbb{V}\mathrm{ar}[Z]/N}}{\ell} . \tag{A.34}$$

The square of the relative error

$$\mathrm{RE}^2 = \frac{\mathbb{V}\mathrm{ar}[\widehat{\ell}]}{\ell^2} \tag{A.35}$$

is called the *squared coefficient of variation*.

EXAMPLE A.3 ***Estimation of Rare-Event Probabilities***

Consider estimation of the tail probability $\ell = \mathbb{P}(X \geq \gamma)$ of some random variable X for a large number γ. If ℓ is very small, then the event $\{X \geq \gamma\}$ is called a *rare event* and the probability $\mathbb{P}(X \geq \gamma)$ is called a *rare-event probability*.

We may attempt to estimate ℓ via (A.33) as

$$\widehat{\ell} = \frac{1}{N} \sum_{i=1}^{N} I_{\{X_i \geq \gamma\}} , \tag{A.36}$$

which involves drawing a random sample $X_1, \ldots, X_N$ from the pdf of X and defining the indicators $Z_i = I_{\{X_i \geq \gamma\}}$, $i = 1, \ldots, N$. The estimator ℓ thus defined is called the *crude Monte Carlo* (CMC) estimator. For small ℓ the relative error of the CMC estimator is given by

$$\mathrm{RE} = \frac{\sqrt{\mathbb{V}\mathrm{ar}[\widehat{\ell}]}}{\mathbb{E}[\widehat{\ell}]} = \sqrt{\frac{1-\ell}{N\,\ell}} \approx \sqrt{\frac{1}{N\,\ell}} . \tag{A.37}$$

As a numerical example, suppose that $\ell = 10^{-6}$. In order to estimate ℓ accurately with relative error, say, $\mathrm{RE} = 0.01$, we need to choose a sample size

$$N \approx \frac{1}{\mathrm{RE}^2 \ell} = 10^{10} .$$

This shows that estimating small probabilities via CMC estimators is computationally meaningless. □

A.3.1 Complexity

The theoretical framework in which one typically examines rare-event probability estimation is based on *complexity theory* (see, for example, [4]).

In particular, the estimators are classified either as *polynomial-time* or as *exponential-time*. It is shown in [4] that for an arbitrary estimator, $\widehat{\ell}$ of ℓ, to be polynomial-time as a function of some γ, it suffices that its squared coefficient of variation, RE^2, or its relative error, RE, is bounded in γ by some polynomial function, $p(\gamma)$. For such polynomial-time estimators, the required sample size to achieve a fixed relative error does not grow too fast as the event becomes rarer.

Consider the estimator (A.36) and assume that ℓ becomes very small as $\gamma \to \infty$. Note that

$$\mathbb{E}[Z^2] \geq (\mathbb{E}[Z])^2 = \ell^2 .$$

Hence, the best one can hope for with such an estimator is that its second moment of Z^2 decreases proportionally to ℓ^2 as $\gamma \to \infty$. We say that the rare-event estimator (A.36) has *bounded relative error* if for all γ

$$\mathbb{E}[Z^2] \leq c\ell^2 \tag{A.38}$$

for some fixed $c \geq 1$. Because bounded relative error is not always easy to achieve, the following weaker criterion is often used. We say that the estimator (A.36) is *logarithmically efficient* (sometimes called *asymptotically optimal*) if

$$\lim_{\gamma\to\infty} \frac{\log \mathbb{E}[Z^2]}{\log \ell^2} = 1 . \tag{A.39}$$

EXAMPLE A.4 ***The CMC Estimator Is Not Logarithmically Efficient***

Consider the CMC estimator (A.36). We have

$$\mathbb{E}[Z^2] = \mathbb{E}[Z] = \ell ,$$

so that

$$\lim_{\gamma\to\infty} \frac{\log \mathbb{E}[Z^2]}{\log \ell^2(\gamma)} = \frac{\log \ell}{\log \ell^2} = \frac{1}{2} .$$

Hence, the CMC estimator is not logarithmically efficient, and therefore alternative estimators must be found to estimate small ℓ. □

A.3.2 Complexity of Randomized Algorithms

A randomized algorithm is said to give an (ε, δ)-*approximation* for a parameter z if its output Z satisfies

$$\mathbb{P}(|Z - z| \leq \varepsilon z) \geq 1 - \delta, \tag{A.40}$$

that is, the "relative error" $|Z - z|/z$ of the approximation Z lies with high probability ($> 1 - \delta$) below some small number ε.

One of the main tools in proving (A.40) for various randomized algorithms is the so-called *Chernoff bound*, which states that for any random variable Y and any number a

$$\mathbb{P}(Y \leq a) \leq \min_{\theta>0} \mathrm{e}^{\theta a}\, \mathbb{E}[\mathrm{e}^{-\theta Y}] . \tag{A.41}$$

Namely, for any fixed a and $\theta > 0$, define the functions $H_1(z) = I_{\{z \leq a\}}$ and $H_2(z) = \mathrm{e}^{\theta(a-z)}$. Then, clearly $H_1(z) \leq H_2(z)$ for all z. As a consequence, for any θ,

$$\mathbb{P}(Y \leq a) = \mathbb{E}[H_1(Y)] \leq \mathbb{E}[H_2(Y)] = \mathrm{e}^{\theta a}\, \mathbb{E}[\mathrm{e}^{-\theta Y}] .$$

The bound (A.41) now follows by taking the smallest such θ.

An important application is the following.

Theorem A.1

Let $X_1, \ldots, X_n$ be iid $\mathsf{Ber}(p)$ random variables, then their sample mean provides an (ε, δ)-approximation for p, that is,

$$\mathbb{P}\left(\left|\frac{1}{n}\sum_{i=1}^{n} X_i - p\right| \leq \varepsilon p\right) \geq 1 - \delta,$$

provided $n \geq 3 \ln (2/\delta)/(p\varepsilon^2)$.

For the proof see, for example, Rubinstein and Kroese [108].

Definition A.1 ***FPRAS***

A randomized algorithm is said to provide a *fully polynomial-time randomized approximation scheme (FPRAS)* if for any input vector $\boldsymbol{x}$ and any parameters $\varepsilon > 0$ and $0 < \delta < 1$ the algorithm outputs an (ε, δ)-approximation to the desired quantity $z(\boldsymbol{x})$ in time, that is, polynomial in ε^{-1}, $\ln\ \delta^{-1}$ and the size n of the input vector $\boldsymbol{x}$.

Note that the sample mean in Theorem A.1 provides a FPRAS for estimating p. Note also that the input vector $\boldsymbol{x}$ consists of the Bernoulli variables $X_1, \ldots, X_n$.

There exists a fundamental connection between the ability to sample *uniformly* from some set $\mathcal{X}$ and counting the number of elements of interest. Because exact uniform sampling is not always feasible, MCMC techniques are often used to sample *approximately* from a uniform distribution.

Definition A.2 ***ε-Uniform Sample***

Let Z be a random output of a sampling algorithm for a finite sample space $\mathcal{X}$. We say that the sampling algorithm generates an *ε-uniform sample* from $\mathcal{X}$ if, for any $\mathcal{Y} \subset \mathcal{X}$,

$$\left|\mathbb{P}(Z \in \mathcal{Y}) - \frac{|\mathcal{Y}|}{|\mathcal{X}|}\right| \leq \varepsilon.$$

Definition A.3 ***Variation Distance***

The variation distance between two distributions F_1 and F_2 on a countable space $\mathcal{X}$ is defined as

$$||F_1 - F_2|| = \frac{1}{2} \sum_{x \in \mathcal{X}} |F_1(x) - F_2(x)|.$$

It is well-known [88] that the definition of variation distance coincides with that of an ε-uniform sample in the sense that a sampling algorithm returns an ε-uniform sample on $\mathcal{X}$ if and only if the variation distance between its output distribution F and the uniform distribution $\mathcal{U}$ satisfies

$$||F - \mathcal{U}|| \leq \varepsilon.$$

Bounding the variation distance between the uniform distribution and the empirical distribution of the Markov chain obtained after some warm-up period is a crucial issue while establishing the foundations of randomized algorithms because, with a bounded variation distance, one can produce an efficient approximation for $|\mathcal{X}^*|$.

Definition A.4 ***FPAUS***

A sampling algorithm is called a *fully polynomial almost uniform sampler (FPAUS)* if, given an input vector $\boldsymbol{x}$ and a parameter $\varepsilon > 0$, the algorithm generates an ε-uniform sample from $\mathcal{X}(\boldsymbol{x})$ and runs in a time that is, polynomial in $\ln \varepsilon^{-1}$ and the size of the input vector $\boldsymbol{x}$.

An important issue is to prove that given an FPAUS for a combinatorial problem, one can construct a corresponding FPRAS.

EXAMPLE A.5 ***FPAUS and FPRAS for Independent Sets***

An FPAUS for independent sets takes as input a graph $G = G(V, E)$ and a parameter $\varepsilon > 0$. The sample space $\mathcal{X}$ consists of all independent sets in G with the output being an ε-uniform sample from $\mathcal{X}$. The time to produce such an ε-uniform sample should be polynomial in the size of the graph and $\ln \varepsilon^{-1}$. Based on the product formula (4.2), Mitzenmacher and Upfal [88] prove that, given an FPAUS, one can construct a corresponding FPRAS. □

Bibliography

[1] E. H. L. Aarts and J. H. M. Korst. *Simulated Annealing and Boltzmann Machines*. Wiley, Chichester, England, 1989.

[2] G. Alon, D. P. Kroese, T. Raviv, and R. Y. Rubinstein. Application of the cross-entropy method to the buffer allocation problem in a simulation-based environment. *Annals of Operations Research*, 134(1):137–151, 2005.

[3] S. Asmussen and P. W. Glynn. *Stochastic Simulation: Algorithms and Analyses*. Springer, New York, 2007.

[4] S. Asmussen and R. Y. Rubinstein. Steady state rare event simulation in queueing model and its complexity properties. In J. Dshalalow, editor, *Advances in Queueing: Theory, Methods and Open Problems*, volume I, pages 429–462. CRC Press, Boca Raton, FL, 1995.

[5] R. Barlow and A. Marshall. Bounds for distribution with monotone hazard rate, I and II. *Annals of Mathematical Statistics*, 35:1234–1274, 1964.

[6] R. Barlow and F. Proschan. *Statistical Theory of Reliability and Life Testing*. Holt, Rinehart & Wilson, New York, 1975.

[7] A. G. Barto and R. S. Sutton. *Reinforcement Learning*. MIT Press, Cambridge, MA, 1998.

[8] A. Ben-Tal and M. Taboule. Penalty functions and duality in stochastic programming via ϕ-divergence functionals. *Mathematics of Operations Reseach*, 12(2):224–240, 1987.

[9] J. Blitzstein and P. Diaconis. A sequential importance sampling algorithm for generating random graphs with prescribed degrees. *Internet Mathematics*, 6:487–520, 2010.

[10] V. S. Borkar and S. P. Meyn. The O.D.E. method for convergence of stochastic approximation and reinforcement learning. *SIAM Journal on Control and Optimization*, 38(2):447–469, 1999.

[11] Z. I. Botev and D. P. Kroese. Global likelihood optimization via the cross-entropy method with an application to mixture models. In *Proceedings of the 36th conference on Winter simulation*, WSC '04, pages 529–535. Winter Simulation Conference, 2004.

[12] Z. I. Botev and D. P. Kroese. An efficient algorithm for rare-event probability estimation, combinatorial optimization, and counting. *Methodology and Computing in Applied Probability*, 10(4):471–505, 2008.

Fast Sequential Monte Carlo Methods for Counting and Optimization, First Edition.
Reuven Y. Rubinstein, Ad Ridder, and Radislav Vaisman.

[13] Z. I. Botev and D. P. Kroese. Efficient Monte Carlo simulation via the generalized splitting method. *Statistics and Computing*, 22:1–16, 2012.

[14] Z. I. Botev, P. L'Ecuyer, G. Rubino, R. Simard, and B. Tuffin. Static network reliability estimation via generalized splitting. *INFORMS Journal on Computing*, 6, 2012.

[15] Z. I. Botev, D. P. Kroese, R. Y. Rubinstein, and P. L'Ecuyer. The cross-entropy method for optimization. In V. Govindaraju and C. R. Rao, editors, *Machine Learning*, volume 31 of Handbook of Statistics. Elsevier, Amsterdam, 2013.

[16] Z. I. Botev, D. P. Kroese, and T. Taimre. Generalized cross-entropy methods. In *Proceedings of RESIM06*, pages 1–30, Bamberg, Germany, October 2006.

[17] A. Boubezoula, S. Paris, and M. Ouladsinea. Application of the cross entropy method to the GLVQ algorithm. *Pattern Recognition*, 41(10):3173–3178, 2008.

[18] Y. Burtin and B. Pittel. Asymptotic estimates of the reliability of a complex system. *Engineering Cybernetics*, 10(3):445–451, 1972.

[19] L. Busoniu, R. Babuska, B. De Schutter, and D. Ernst. *Reinforcement Learning and Dynamic Programming Using Function Approximators*. CRC Press, Boca Raton, FL, 2010.

[20] H. Cancela, M. El Khadiri, and G. Rubino. Rare event analysis by Monte Carlo techniques in static models. In G. Rubino and B. Tuffin, editors, *Rare Event Simulation Using Monte Carlo Methods*, pages 145–170. Wiley, Chichester, England, 2009. Chapter 7.

[21] H. Cancela and M. El Khadiri. On the RVR simulation algorithm for network reliability evaluation. *IEEE Transactions on Reliability*, 52(2):207–212, 2003.

[22] H. Cancela, P. L'Ecuyer, M. Lee, G. Rubino, and B. Tuffin. Analysis and improvements of path-based methods for Monte Carlo reliability evaluation of static models. In J. Faulin, A. A. Juan, S. Martorell, and E. Ramirez-Marquez, editors, *Simulation Methods for Reliability and Availability of Complex Systems*, pages 65–84. Springer, London, 2010.

[23] H. Cancela, P. L'Ecuyer, G. Rubino, and B. Tuffin. Combination of conditional Monte Carlo and approximate zero-variance importance sampling for network reliability estimation. In *Proceedings of the 42nd Winter Simulation Conference*, WSC '10, pages 1263–1274. Winter Simulation Conference, 2010.

[24] F. Cérou and A. Guyader. Adaptive multilevel splitting for rare event analysis. *Stochastic Analysis and Applications*, 25(2):417–443, 2007.

[25] J. C. C. Chan and D. P. Kroese. Improved cross-entropy method for estimation. *Statistics and Computing*, 22(5):1031–1040, 2012.

[26] G. M. J-B. Chaslot, M. H. M. Winands, I. Szita, and H. J. van den Herik. Cross-entropy for Monte-Carlo tree search. *ICGA Journal*, 31(3):145–156, 2008.

[27] J.-C. Chen, C.-K. Wen, C.-P. Li, and P. Ting. Cross-entropy optimization for the design of fiber Bragg gratings. *Photonics Journal, IEEE*, 4(5):1495–1503, oct. 2012.

[28] Y. Chen, P. Diaconis, S. P. Holmes, and J. Liu. Sequential Monte Carlo methods for statistical analysis of tables. *Journal of the American Statistical Association*, 100:109–120, 2005.

[29] K. Chepuri and T. Homem-de-Mello. Solving the vehicle routing problem with stochastic demands using the cross entropy method. *Annals of Operations Research*, 134(1):153–181, 2005.

[30] N. Clisby. Efficient implementation of the pivot algorithm for self-avoiding walks. *Journal of Statistical Physics*, 140:349–392, 2010.

[31] I. Cohen, B. Golany, and A. Shtub. Resource allocation in stochastic, finite-capacity, multi-project systems through the cross entropy methodology. *Journal of Scheduling*, 10(1):181–193, 2007.

[32] T. H. Cormen, C. E. Leiserson, and R. L. Rivest. *Introduction to Algorithms*. MIT Press, Cambridge, MA, third edition, 2009.

[33] A. Costa, J. Owen, and D. P. Kroese. Convergence properties of the cross-entropy method for discrete optimization. *Operations Research Letters*, 35(5):573–580, 2007.

[34] T. M. Cover and J. A. Thomas. *Elements of Information Theory*. Wiley, Hoboken, NJ, second edition, 2006.

[35] M. Davis, G. Logemann, and D. Loveland. A machine program for theorem proving. *Communications of the ACM,* 5:394–397, 1962.

[36] M. Davis and H. Putnam. A computing procedure for quantification theory. *Journal of the ACM*, 7:201–215, 1960.

[37] P. T. de Boer, D. P. Kroese, S. Mannor, and R. Y. Rubinstein. A tutorial on the cross-entropy method. *Annals of Operations Research*, 134(1):19–67, 2005.

[38] T. Dean and P. Dupuis. Splitting for rare event simulation: A large deviations approach to design and analysis. *Stochastic Processes and their Applications*, 119(2):562–587, 2009.

[39] T. Dean and P. Dupuis. The design and analysis of a generalized RESTART/DPR algorithm for rare event simulation. *Annals of Operations Research*, 189(1):63–102, 2011.

[40] P. Dupuis, B. Kaynar, R. Reuvenstein, A. Ridder, and R. Vaisman. Counting with combined splitting and capture-recapture methods. *Stochastic Models*, 28:478–502, 2012.

[41] T. Elperin, I. B. Gertsbakh, and M. Lomonosov. Estimation of network reliability using graph evolution models. *IEEE Transactions on Reliability*, 40(5):572–581, 1991.

[42] T. Elperin, I. B. Gertsbakh, and M. Lomonosov. An evolution model for Monte Carlo estimation of equilibrium network renewal parameters. *Probability in the Engineering and Informational Sciences*, 6:457–469, 1992.

[43] T. Elperin, I. B. Gertsbakh, and M. Lomonosov. On Monte Carlo estimates in network reliability. *Probability in the Engineering and Informational Sciences*, 8:245–265, 1994.

[44] P. Erdös and A. Rényi. On the evolution of random graphs. In *Publication of the Mathematical Institute of the Hungarian Academy of Sciences*, pages 17–61, 1960.

[45] D. Ernst, M. Glavic, G.-B. Stan, S. Mannor, and L. Wehenkel. The cross-entropy method for power system combinatorial optimization problems. In *Proceedings of the 7th IEEE Power Engineering Society (IEEE-PowerTech 2007)*, pages 1290–1295, 2007.

[46] G. E. Evans, J. M. Keith, and D. P. Kroese. Parallel cross-entropy optimization. In *Proceedings of the 39th conference on Winter simulation*, WSC '07, pages 2196–2202, Winter Simulation Conference, 2007.

[47] G. Fishman. A Monte Carlo sampling plan for estimating network reliability. *Operations Research*, 34(4):581–594, 1980.

[48] M. J. J. Garvels. *The splitting method in rare-event simulation*. PhD thesis, University of Twente, 2000.

[49] M. J. J. Garvels and D. P. Kroese. A comparison of RESTART implementations. In *Proceedings of the 30th conference on Winter simulation*, WSC '98, pages 601–608, Winter Simulation Conference, 1998.

[50] M. J. J. Garvels, D. P. Kroese, and J. C. W. van Ommeren. On the importance function RESTART simulation. *European Transactions on Telecommunications*, 13(4), 2002.

[51] A. Gelman. and D. B. Rubin. Inference from iterative simulation using multiple sequences (with discussion). *Statistical Science*, 7:457–511, 1992.

[52] I. B. Gertsbakh and Y. Shpungin. *Models of Network Reliability: Analysis, Combinatorics, and Monte Carlo*. CRC Press, Boca Raton, FL, 2010.

[53] A. Ghate and R. L. Smith. A dynamic programming approach to efficient sampling from Boltzmann distribution. *Operations Research Letters*, 36(6):665–668, 2008.

[54] P. Glasserman, P. Heidelberger, P. Shahabuddin, and T. Zajic. A large deviations perspective on the efficiency of multilevel splitting. *IEEE Transactions on Automatic Control*, 43(12):1666–1679, 1998.

[55] P. Glasserman, P. Heidelberger, P. Shahabuddin, and T. Zajic. Multilevel splitting for estimating rare event probabilities. *Operations Research*, 47(4):585–600, 1999.

[56] V. Gogate and R. Dechter. Approximate counting by sampling the backtrack-free search space. In *Proceedings of the 22nd national conference on Artificial intelligence—Volume 1,* AAAI'07, pages 198–203. AAAI Press, Menlo Park, CA, 2007.

[57] V. Gogate and R. Dechter. Samplesearch: Importance sampling in presence of determinism. *Artificial Intelligence*, 175(2):694–729, 2011.

[58] E. Gryazina and B. Polyak. Randomized methods based on new Monte Carlo schemes for control and optimizations. *Annals of Operations Research*, 189(1):343–356, 2011.

[59] K. P. Hui, N. Bean, M. Kraetzl, and D. P. Kroese. The cross-entropy method for network reliability estimation. *Annals of Operations Research*, 134(1):101–118, 2005.

[60] M. Jerrum, L. G. Valiant, and V. V. Vazirani. Random generation of combinatorial structures from a uniform distribution. *Theoretical Computer Science*, 43:169–188, 1986.

[61] L. P. Kaelbling, M. Littman, and A. W. Moore. Reinforcement learning—A survey. *Journal of Artificial Intelligence Research*, 4:237–285, May 1996.

[62] H. Kahn and T. E. Harris. Estimation of particle transmission by random sampling. *National Bureau of Standards Appl. Math. Series*, 12:27–30, 1951.

[63] J. N. Kapur and H. K. Kesavan. *Entropy Optimization with Applications*. Academic Press, Boston, 1992.

[64] D. R. Karger. A randomized fully polynomial time approximation scheme for the all terminal network reliability problem. In *Proceedings of the 27th annual ACM symposium on Theory of computing*, STOC '95, pages 11–17, New York, NY, USA, 1995. ACM.

[65] R. M. Karp, M. Luby, and N. Madras. Monte-Carlo approximation algorithms for enumeration problems. *Journal of Algorithms*, 10(3):429–448, 1989.

[66] M. Kearns and S. Singh. Near-optimal reinforcement learning in polynomial time. *Machine Learning*, 49(2-3):209–232, 2002.

[67] J. Keith and D. P. Kroese. Rare event simulation and combinatorial optimization using cross entropy: Sequence alignment by rare event simulation. In *Proceedings*

of the 34th conference on Winter simulation, WSC '02, pages 320–327. Winter Simulation Conference, 2002.

[68] D. E. Knuth. *The Art of Computer Programming, Combinatorial Algorithms, Part 1*, volume 4A. Addison-Wesley, Upper Saddle River, NJ, 2011.

[69] R. P. Kothari and D. P. Kroese. Optimal generation expansion planning via the cross-entropy method. In *Proceedings of the 41st conference on Winter simulation*, WSC '09, pages 1482–1491. Winter Simulation Conference, 2009.

[70] D. P. Kroese, K.-P. Hui, and S. Nariai. Network reliability optimization via the cross-entropy method. *IEEE Transactions on Reliability*, 56(2):275–287, 2007.

[71] D. P. Kroese, S. Porotsky, and R. Y. Rubinstein. The cross-entropy method for continuous multi-extremal optimization. *Methodology and Computing in Applied Probability*, 8(3):383–407, 2006.

[72] D. P. Kroese, R. Y. Rubinstein, and T. Taimre. Application of the cross-entropy method to clustering and vector quantization. *Journal of Global Optimization*, 37:137–157, 2007.

[73] D. P. Kroese, T. Taimre, and Z. I. Botev. *Handbook of Monte Carlo Methods*. Wiley, New Jersey, 2011.

[74] H. W. Kuhn. The Hungarian method for the assignment problem. *Naval Research Logistics Quarterly*, 2:83–97, 1955.

[75] H. W. Kuhn. Nonlinear programming: a historical view. *ACM SIGMAP Bulletin*, 31:6–18, June 1982.

[76] P. L'Ecuyer, V. Demeres, and B. Tuffin. Splitting for rare-event simulation. In *Proceedings of the 38th conference on Winter simulation*, WSC '06, pages 137–148. Winter Simulation Conference, 2006.

[77] P. L'Ecuyer, V. Demeres, and B. Tuffin. Rare-events, splitting, and quasi-Monte Carlo. *ACM Transactions on Modeling and Computer Simulation*, 12(2), 2007.

[78] M. Liskiewicz, M. Ogiharaba, and S. Toda. The complexity of counting self-avoiding walks in subgraphs of two-dimensional grids and hypercubes. *Theoretical Computer Science*, 304:129–156, 2003.

[79] J. S. Liu. *Monte Carlo Strategies in Scientific Computing*. Springer, New York, 2001.

[80] Z. Liu, A. Doucet, and S. S. Singh. The cross-entropy method for blind multiuser detection. In *Proceedings 2004 IEEE International Symposium on Information Theory*, page 510, Chicago, IL, 2004.

[81] A. Lörincza, Z. Palotaia, and G. Szirtesb. Spike-based cross-entropy method for reconstruction. *Neurocomputing*, 71(16-18):3635–3639, 2008.

[82] N. Madras and A. D. Sokal. The pivot algorithm: A highly efficient Monte Carlo method for the self-avoiding walk. *Journal of Statistical Physics*, 50(1):109–186, 1988.

[83] S. Mannor, R. Y. Rubinstein, and Y. Gat. The cross-entropy method for fast policy search. In *Proceedings of the 20th International Conference on Machine Learning*, pages 512–519, AAAI Press, Menlo Park, CA, 2003.

[84] L. Margolin. On the convergence of the cross-entropy method. *Annals of Operations Research*, 134(1):201–214, 2005.

[85] V. B. Melas. On the efficiency of the splitting and roulette approach for sensitivity analysis. In *Proceedings of the 29th conference on Winter simulation*, WSC '97, pages 269–274, Winter Simulation Conference, 1997.

[86] I. Menache and S. Mannor. Basis function adaptation in temporal difference reinforcement learning. *Annals of Operations Research*, 134(1):215–238, 2005.

[87] P. Metzner, C. Schutte, and E. Vanden-Eijnden. Illustration of transition path theory on a collection of simple examples. *The Journal of Chemical Physics*, 125(8):084–110, 2006.

[88] M. Mitzenmacher and E. Upfal. *Probability and Computing: Randomized Algorithms and Probabilistic Analysis*. Cambridge University Press, Cambridge, 2005.

[89] P. Del Moral. *Feynman-Kac Formulae: Genealogical and Interacting Particle Systems with Applications. Probability and Its Applications*. Springer, New York, 2004.

[90] R. Motwani and R. Raghavan. *Randomized Algorithms*. Cambridge University Press, Cambridge, 1995.

[91] M. S. Nikulin. Chi-squared test for normality. *Proceedings of the International Vilnius Conference on Probability Theory and Mathematical Statistics*, Volume 2:119–122, 1973.

[92] C. H. Papadimitriou and K. Steiglitz. *Combinatorial Optimization: Algorithms and Complexity*. Prentice Hall, Englewood Cliffs, NJ, 1982.

[93] V. Pihur, S. Datta, and S. Datta. Weighted rank aggregation of cluster validation measures: A Monte Carlo cross-entropy approach. *Bioinformatics*, 23(13):1607–1615, 2007.

[94] M. A. Pincus. A closed form selection of certain programming problems. *Operations Research*, 16:690–694, 1968.

[95] M. Puterman. *Markov Decision Processes*. Wiley, New York, 1994.

[96] B. Roberts and D. P. Kroese. Estimating the number of s-t paths in a graph. *Journal of Graph Algorithms and Applications*, 11(1):195–214, 2007.

[97] M. N. Rosenbluth and A. W. Rosenbluth. Monte Carlo calculation of the average extension of molecular chains. *Journal of Chemical Physics*, 23(2):356–359, 1955.

[98] M. T. Rosenstein and A. G. Barto. Robot weightlifting by direct policy search. In *Proceedings of the 17th International Joint Conference on Artificial Intelligence—Volume 2*, IJCAI'01, pages 839–846, Morgan Kaufmann Publishers, San Francisco (CA), 2001.

[99] S. M. Ross. *Simulation*. Elsevier Academic Press, Amsterdam, fourth edition, 2006.

[100] R. Y. Rubinstein. Optimization of computer simulation models with rare events. *European Journal of Operational Research*, 99(1):89–112, 1997.

[101] R. Y. Rubinstein. The cross-entropy method for combinatorial and continuous optimization. *Methodology and Computing in Applied Probability*, 1(2):127–190, 1999.

[102] R. Y. Rubinstein. Combinatorial optimization, cross-entropy, ants and rare events. In S. Uryasev and P. M. Pardalos, editors, *Stochastic Optimization: Algorithms and Applications*, pages 304–358. Kluwer, Dordrecht, 2001.

[103] R. Y. Rubinstein. A stochastic minimum cross-entropy method for combinatorial optimization and rare-event estimation. *Methodology and Computing in Applied Probability*, 7(1):1–46, 2005.

[104] R. Y. Rubinstein. The Gibbs cloner for combinatorial optimization, counting and sampling. *Methodology and Computing in Applied Probability*, 11(2):491–549, 2009.

[105] R. Y. Rubinstein. Randomized algorithms with splitting: Why the classic randomized algorithms do not work and how to make them work. *Methodology and Computing in Applied Probability*, 12(1):1–41, 2010.

[106] R. Y. Rubinstein and P. W. Glynn. How to deal with the curse of dimensionality of likelihood ratios in Monte Carlo simulations. *Stochastic Models*, 25(4):547–568, 2009.

[107] R. Y. Rubinstein and D. P. Kroese. *The Cross-Entropy Method: A Unified Approach to Combinatorial Optimization, Monte-Carlo Simulation and Machine Learning*. Springer, New York, 2004.

[108] R. Y. Rubinstein and D. P. Kroese. *Simulation and the Monte Carlo Method*. Wiley, Hoboken, NJ, second edition, 2008.

[109] R. Y. Rubinstein, Z. Botev, and R. Vaisman. Hanging edges for fast reliability estimation. Private Communication, 2012.

[110] F. J. Samaniego. On closure of the IFR under formation of coherent systems. *IEEE Transactions on Reliability*, 34:69–72, 1985.

[111] A. Sani. *Stochastic Modelling and Intervention of the Spread of HIV/AIDS*. PhD thesis, The University of Queensland, Brisbane, 2009.

[112] A. Sani and D. P. Kroese. Controlling the number of HIV infectives in a mobile population. *Mathematical Biosciences*, 213(2):103–112, 2008.

[113] G. A. F. Seber. The effect of trap response on tag recapture estimates. *Biometrics*, 26:13–22, 1970.

[114] G. A. F. Seber. *Estimation of Animal Abundance and Related Parameters*. Blackburn Press, Caldwell, NJ, second edition, 2002.

[115] D. Siegmund. Importance sampling in the Monte Carlo study of sequential tests. *Annals of Statististics*, 4:673–684, 1976.

[116] Z. Szabó, B. Póczos, and A. Lörinc. Cross-entropy optimization for independent process analysis. In *Independent Component Analysis and Blind Signal Separation, Lecture Notes in Computer Science*, Volume 3889, pages 909–916, Springer, Heidelberg, 2006.

[117] J. N. Tsitsiklis. Asynchronous stochastic approximation and Q-learning. *Machine Learning*, 16:185–202, 1994.

[118] A. Ünveren and A. Acan. Multi-objective optimization with cross entropy method: Stochastic learning with clustered pareto fronts. In *Proceedings of the IEEE Congress on Evolutionary Computation,* pages 3065–3071, 2007.

[119] L. G. Valiant. The complexity of computing the permanent. *Theoretical Computer Science*, 8:189–201, 1979.

[120] L. G. Valiant. The complexity of enumeration and reliability problems. *SIAM Journal on Computing*, 8:410–421, 1979.

[121] E. J. J. van Rensburg. Monte Carlo methods for the self-avoiding walk. *Journal of Physics A: Mathematical and Theoretical*, 42(32):323001, 2009.

[122] M. Villén-Altimirano and J. Villén-Altimirano. RESTART: A method for accelerating rare event simulations. In *Proceedings of the 13th International Teletraffic Congress, Performance and Control in ATM,* pages 71–76, June 1991.

[123] M. Villén-Altimirano and J. Villén-Altimirano. RESTART: A straightforward method for fast simulation of rare events. In *Proceedings of the 26th conference on Winter simulation*, WSC '94, pages 282–289, Winter Simulation Conference, 1994.

[124] M. Villén-Altimirano and J. Villén-Altimirano. About the efficiency of RESTART. In *Proceedings of the 1999 RESIM Workshop*, pages 99–128, University of Twente, the Netherlands, 1999.

[125] C. Watkins. *Learning from Delayed Rewards.* PhD thesis, Cambridge University, 1989.

[126] W. Wei and B. Selman. A new approach to model counting. In *Proceedings of the 8th international conference on Theory and Applications of Satisfiability Testing, SAT'05, Lecture Notes in Computer Science*, Volume 3569, pages 324–339, Springer, Heidelberg, 2005.

[127] R. J. Williams. Simple statistical gradient-following algorithms for connectionist reinforcement learning. *Machine Learning*, 8(3-4,):229–256, 1992.

[128] Y. Wu and C. Fyfe. Topology perserving mappings using cross entropy adaptation. In *Proceedings of the 7th WSEAS International Conference on Artificial intelligence, knowledge engineering and data bases*, AIKED'08, pages 176–181, World Scientific and Engineering Academy and Society, Stevens Point, WI, 2008.

Abbreviations and Acronyms

BMC	basic Monte Carlo
BME	basic MinxEnt
cdf	cumulative distribution function
CAP-RECAP	capture-recapture
CE	cross-entropy
CMC	crude Monte Carlo
CNF	conjunctive normal form
DNF	disjunctive normal form
ECM	exponential change of measure
FPAUS	fully polynomial almost uniform sampler
FPRAS	fully polynomial randomized approximation scheme
HMC	hanging edges Monte Carlo
IME	indicator MinxEnt
iid	independent and identically distributed
MCMC	Markov chain Monte Carlo
MDP	Markov decision process
MinxEnt	minimum cross-entropy
OSLA	one-step-look-ahead
nSLA	n-step-look-ahead
pdf	probability density function (both discrete and continuous)
PMC	permutation Monte Carlo
RE	relative error
RL	reinforced learning
SA	stochastic approximation
SAT	satisfiability (problem)
SAW	self-avoiding walk
SE	stochastic enumeration
SIS	sequential importance sampling
SMC	sequential Monte Carlo
TSP	traveling salesman problem

Fast Sequential Monte Carlo Methods for Counting and Optimization, First Edition.
Reuven Y. Rubinstein, Ad Ridder, and Radislav Vaisman.

List of Symbols

$\gg$	much greater than
$\propto$	proportional to
$\sim$	is distributed according to
$\stackrel{\mathcal{D}}{=}$	equal in distribution
$\approx$	approximately
$\mathbb{E}$	expectation
$\mathbb{N}$	set of natural numbers $\{0, 1, \ldots\}$
$\mathbb{P}$	probability measure
$\mathbb{R}$	the real line = one-dimensional Euclidean space
$\mathbb{R}^n$	n-dimensional Euclidean space
$\mathcal{D}$	Kullback-Leibler divergence
$\mathcal{S}$	Shannon entropy
Ber	Bernoulli distribution
Bin	binomial distribution
Exp	exponential distribution
N	normal or Gaussian distribution
U	uniform distribution
α	smoothing parameter or acceptance probability
γ	level parameter
ρ	rarity parameter
f	probability density (discrete or continuous)
g	importance sampling density
I_A	indicator function of event A
log	(natural) logarithm
N	sample size
$N^{(e)}$	elite sample size
$N^{(s)}$	screened elite sample size

Fast Sequential Monte Carlo Methods for Counting and Optimization, First Edition.
Reuven Y. Rubinstein, Ad Ridder, and Radislav Vaisman.

$\mathcal{O}$	Big-O order symbol
S	performance function
$S_{(i)}$	i-th order statistic
$\boldsymbol{u}$	nominal reference parameter (vector)
$\boldsymbol{v}$	reference parameter (vector)
$\widehat{\boldsymbol{v}}$	estimated reference parameter
W	likelihood ratio
$\boldsymbol{x}, \boldsymbol{y}$	vectors
$\boldsymbol{X}, \boldsymbol{Y}$	random vectors
$\mathcal{X}$	sets

Index

Fast Sequential Monte Carlo Methods for Counting and Optimization, First Edition.
Reuven Y. Rubinstein, Ad Ridder, and Radislav Vaisman.

WILEY SERIES IN PROBABILITY AND STATISTICS

ESTABLISHED BY WALTER A. SHEWHART AND SAMUEL S. WILKS

Editors: *David J. Balding, Noel A. C. Cressie, Garrett M. Fitzmaurice, Harvey Goldstein, Iain M. Johnstone, Geert Molenberghs, David W. Scott, Adrian F. M. Smith, Ruey S. Tsay, Sanford Weisberg*

Editors Emeriti: *Vic Barnett, J. Stuart Hunter, Joseph B. Kadane, Jozef L. Teugels*

The ***Wiley Series in Probability and Statistics*** is well established and authoritative. It covers many topics of current research interest in both pure and applied statistics and probability theory. Written by leading statisticians and institutions, the titles span both state-of-the-art developments in the field and classical methods.

Reflecting the wide range of current research in statistics, the series encompasses applied, methodological and theoretical statistics, ranging from applications and new techniques made possible by advances in computerized practice to rigorous treatment of theoretical approaches.

This series provides essential and invaluable reading for all statisticians, whether in academia, industry, government, or research.

† ABRAHAM and LEDOLTER · Statistical Methods for Forecasting
AGRESTI · Analysis of Ordinal Categorical Data, *Second Edition*
AGRESTI · An Introduction to Categorical Data Analysis, *Second Edition*
AGRESTI · Categorical Data Analysis, *Second Edition*
ALTMAN, GILL, and McDONALD · Numerical Issues in Statistical Computing for the Social Scientist
AMARATUNGA and CABRERA · Exploration and Analysis of DNA Microarray and Protein Array Data
ANDĚL · Mathematics of Chance
ANDERSON · An Introduction to Multivariate Statistical Analysis, *Third Edition*
* ANDERSON · The Statistical Analysis of Time Series
ANDERSON, AUQUIER, HAUCK, OAKES, VANDAELE, and WEISBERG · Statistical Methods for Comparative Studies
ANDERSON and LOYNES · The Teaching of Practical Statistics
ARMITAGE and DAVID (editors) · Advances in Biometry
ARNOLD, BALAKRISHNAN, and NAGARAJA · Records
* ARTHANARI and DODGE · Mathematical Programming in Statistics
* BAILEY · The Elements of Stochastic Processes with Applications to the Natural Sciences
BAJORSKI · Statistics for Imaging, Optics, and Photonics
BALAKRISHNAN and KOUTRAS · Runs and Scans with Applications
BALAKRISHNAN and NG · Precedence-Type Tests and Applications
BARNETT · Comparative Statistical Inference, *Third Edition*
BARNETT · Environmental Statistics
BARNETT and LEWIS · Outliers in Statistical Data, *Third Edition*
BARTHOLOMEW, KNOTT, and MOUSTAKI · Latent Variable Models and Factor Analysis: A Unified Approach, *Third Edition*
BARTOSZYNSKI and NIEWIADOMSKA-BUGAJ · Probability and Statistical Inference, *Second Edition*
BASILEVSKY · Statistical Factor Analysis and Related Methods: Theory and Applications
BATES and WATTS · Nonlinear Regression Analysis and Its Applications
BECHHOFER, SANTNER, and GOLDSMAN · Design and Analysis of Experiments for Statistical Selection, Screening, and Multiple Comparisons

*Now available in a lower priced paperback edition in the Wiley Classics Library.
†Now available in a lower priced paperback edition in the Wiley–Interscience Paperback Series.

BEIRLANT, GOEGEBEUR, SEGERS, TEUGELS, and DE WAAL · Statistics of Extremes: Theory and Applications
BELSLEY · Conditioning Diagnostics: Collinearity and Weak Data in Regression
† BELSLEY, KUH, and WELSCH · Regression Diagnostics: Identifying Influential Data and Sources of Collinearity
BENDAT and PIERSOL · Random Data: Analysis and Measurement Procedures, *Fourth Edition*
BERNARDO and SMITH · Bayesian Theory
BHAT and MILLER · Elements of Applied Stochastic Processes, *Third Edition*
BHATTACHARYA and WAYMIRE · Stochastic Processes with Applications
BIEMER, GROVES, LYBERG, MATHIOWETZ, and SUDMAN · Measurement Errors in Surveys
BILLINGSLEY · Convergence of Probability Measures, *Second Edition*
BILLINGSLEY · Probability and Measure, *Anniversary Edition*
BIRKES and DODGE · Alternative Methods of Regression
BISGAARD and KULAHCI · Time Series Analysis and Forecasting by Example
BISWAS, DATTA, FINE, and SEGAL · Statistical Advances in the Biomedical Sciences: Clinical Trials, Epidemiology, Survival Analysis, and Bioinformatics
BLISCHKE and MURTHY (editors) · Case Studies in Reliability and Maintenance
BLISCHKE and MURTHY · Reliability: Modeling, Prediction, and Optimization
BLOOMFIELD · Fourier Analysis of Time Series: An Introduction, *Second Edition*
BOLLEN · Structural Equations with Latent Variables
BOLLEN and CURRAN · Latent Curve Models: A Structural Equation Perspective
BOROVKOV · Ergodicity and Stability of Stochastic Processes
BOSQ and BLANKE · Inference and Prediction in Large Dimensions
BOULEAU · Numerical Methods for Stochastic Processes
* BOX and TIAO · Bayesian Inference in Statistical Analysis
BOX · Improving Almost Anything, *Revised Edition*
* BOX and DRAPER · Evolutionary Operation: A Statistical Method for Process Improvement
BOX and DRAPER · Response Surfaces, Mixtures, and Ridge Analyses, *Second Edition*
BOX, HUNTER, and HUNTER · Statistics for Experimenters: Design, Innovation, and Discovery, *Second Editon*
BOX, JENKINS, and REINSEL · Time Series Analysis: Forcasting and Control, *Fourth Edition*
BOX, LUCEÑO, and PANIAGUA-QUIÑONES · Statistical Control by Monitoring and Adjustment, *Second Edition*
* BROWN and HOLLANDER · Statistics: A Biomedical Introduction
CAIROLI and DALANG · Sequential Stochastic Optimization
CASTILLO, HADI, BALAKRISHNAN, and SARABIA · Extreme Value and Related Models with Applications in Engineering and Science
CHAN · Time Series: Applications to Finance with R and S-Plus®, *Second Edition*
CHARALAMBIDES · Combinatorial Methods in Discrete Distributions
CHATTERJEE and HADI · Regression Analysis by Example, *Fourth Edition*
CHATTERJEE and HADI · Sensitivity Analysis in Linear Regression
CHERNICK · Bootstrap Methods: A Guide for Practitioners and Researchers, *Second Edition*
CHERNICK and FRIIS · Introductory Biostatistics for the Health Sciences
CHILÈS and DELFINER · Geostatistics: Modeling Spatial Uncertainty, *Second Edition*
CHOW and LIU · Design and Analysis of Clinical Trials: Concepts and Methodologies, *Second Edition*
CLARKE · Linear Models: The Theory and Application of Analysis of Variance
CLARKE and DISNEY · Probability and Random Processes: A First Course with Applications, *Second Edition*

*Now available in a lower priced paperback edition in the Wiley Classics Library.
†Now available in a lower priced paperback edition in the Wiley–Interscience Paperback Series.

* COCHRAN and COX · Experimental Designs, *Second Edition*
COLLINS and LANZA · Latent Class and Latent Transition Analysis: With Applications in the Social, Behavioral, and Health Sciences
CONGDON · Applied Bayesian Modelling
CONGDON · Bayesian Models for Categorical Data
CONGDON · Bayesian Statistical Modelling, *Second Edition*
CONOVER · Practical Nonparametric Statistics, *Third Edition*
COOK · Regression Graphics
COOK and WEISBERG · An Introduction to Regression Graphics
COOK and WEISBERG · Applied Regression Including Computing and Graphics
CORNELL · A Primer on Experiments with Mixtures
CORNELL · Experiments with Mixtures, Designs, Models, and the Analysis of Mixture Data, *Third Edition*
COX · A Handbook of Introductory Statistical Methods
CRESSIE · Statistics for Spatial Data, *Revised Edition*
CRESSIE and WIKLE · Statistics for Spatio-Temporal Data
CSÖRGŐ and HORVÁTH · Limit Theorems in Change Point Analysis
DAGPUNAR · Simulation and Monte Carlo: With Applications in Finance and MCMC
DANIEL · Applications of Statistics to Industrial Experimentation
DANIEL · Biostatistics: A Foundation for Analysis in the Health Sciences, *Eighth Edition*
* DANIEL · Fitting Equations to Data: Computer Analysis of Multifactor Data, *Second Edition*
DASU and JOHNSON · Exploratory Data Mining and Data Cleaning
DAVID and NAGARAJA · Order Statistics, *Third Edition*
* DEGROOT, FIENBERG, and KADANE · Statistics and the Law
DEL CASTILLO · Statistical Process Adjustment for Quality Control
DEMARIS · Regression with Social Data: Modeling Continuous and Limited Response Variables
DEMIDENKO · Mixed Models: Theory and Applications with R, *Second Edition*
DENISON, HOLMES, MALLICK and SMITH · Bayesian Methods for Nonlinear Classification and Regression
DETTE and STUDDEN · The Theory of Canonical Moments with Applications in Statistics, Probability, and Analysis
DEY and MUKERJEE · Fractional Factorial Plans
DILLON and GOLDSTEIN · Multivariate Analysis: Methods and Applications
* DODGE and ROMIG · Sampling Inspection Tables, *Second Edition*
* DOOB · Stochastic Processes
DOWDY, WEARDEN, and CHILKO · Statistics for Research, *Third Edition*
DRAPER and SMITH · Applied Regression Analysis, *Third Edition*
DRYDEN and MARDIA · Statistical Shape Analysis
DUDEWICZ and MISHRA · Modern Mathematical Statistics
DUNN and CLARK · Basic Statistics: A Primer for the Biomedical Sciences, *Fourth Edition*
DUPUIS and ELLIS · A Weak Convergence Approach to the Theory of Large Deviations
EDLER and KITSOS · Recent Advances in Quantitative Methods in Cancer and Human Health Risk Assessment
* ELANDT-JOHNSON and JOHNSON · Survival Models and Data Analysis
ENDERS · Applied Econometric Time Series, *Third Edition*
† ETHIER and KURTZ · Markov Processes: Characterization and Convergence
EVANS, HASTINGS, and PEACOCK · Statistical Distributions, *Third Edition*
EVERITT, LANDAU, LEESE, and STAHL · Cluster Analysis, *Fifth Edition*
FEDERER and KING · Variations on Split Plot and Split Block Experiment Designs

*Now available in a lower priced paperback edition in the Wiley Classics Library.
†Now available in a lower priced paperback edition in the Wiley–Interscience Paperback Series.

FELLER · An Introduction to Probability Theory and Its Applications, Volume I, *Third Edition,* Revised; Volume II, *Second Edition*
FITZMAURICE, LAIRD, and WARE · Applied Longitudinal Analysis, *Second Edition*
* FLEISS · The Design and Analysis of Clinical Experiments
FLEISS · Statistical Methods for Rates and Proportions, *Third Edition*
† FLEMING and HARRINGTON · Counting Processes and Survival Analysis
FUJIKOSHI, ULYANOV, and SHIMIZU · Multivariate Statistics: High-Dimensional and Large-Sample Approximations
FULLER · Introduction to Statistical Time Series, *Second Edition*
† FULLER · Measurement Error Models
GALLANT · Nonlinear Statistical Models
GEISSER · Modes of Parametric Statistical Inference
GELMAN and MENG · Applied Bayesian Modeling and Causal Inference from ncomplete-Data Perspectives
GEWEKE · Contemporary Bayesian Econometrics and Statistics
GHOSH, MUKHOPADHYAY, and SEN · Sequential Estimation
GIESBRECHT and GUMPERTZ · Planning, Construction, and Statistical Analysis of Comparative Experiments
GIFI · Nonlinear Multivariate Analysis
GIVENS and HOETING · Computational Statistics
GLASSERMAN and YAO · Monotone Structure in Discrete-Event Systems
GNANADESIKAN · Methods for Statistical Data Analysis of Multivariate Observations, *Second Edition*
GOLDSTEIN · Multilevel Statistical Models, *Fourth Edition*
GOLDSTEIN and LEWIS · Assessment: Problems, Development, and Statistical Issues
GOLDSTEIN and WOOFF · Bayes Linear Statistics
GREENWOOD and NIKULIN · A Guide to Chi-Squared Testing
GROSS, SHORTLE, THOMPSON, and HARRIS · Fundamentals of Queueing Theory, *Fourth Edition*
GROSS, SHORTLE, THOMPSON, and HARRIS · Solutions Manual to Accompany Fundamentals of Queueing Theory, *Fourth Edition*
* HAHN and SHAPIRO · Statistical Models in Engineering
HAHN and MEEKER · Statistical Intervals: A Guide for Practitioners
HALD · A History of Probability and Statistics and their Applications Before 1750
† HAMPEL · Robust Statistics: The Approach Based on Influence Functions
HARTUNG, KNAPP, and SINHA · Statistical Meta-Analysis with Applications
HEIBERGER · Computation for the Analysis of Designed Experiments
HEDAYAT and SINHA · Design and Inference in Finite Population Sampling
HEDEKER and GIBBONS · Longitudinal Data Analysis
HELLER · MACSYMA for Statisticians
HERITIER, CANTONI, COPT, and VICTORIA-FESER · Robust Methods in Biostatistics
HINKELMANN and KEMPTHORNE · Design and Analysis of Experiments, Volume 1: Introduction to Experimental Design, *Second Edition*
HINKELMANN and KEMPTHORNE · Design and Analysis of Experiments, Volume 2: Advanced Experimental Design
HINKELMANN (editor) · Design and Analysis of Experiments, Volume 3: Special Designs and Applications
HOAGLIN, MOSTELLER, and TUKEY · Fundamentals of Exploratory Analysis of Variance
* HOAGLIN, MOSTELLER, and TUKEY · Exploring Data Tables, Trends and Shapes
* HOAGLIN, MOSTELLER, and TUKEY · Understanding Robust and Exploratory Data Analysis

*Now available in a lower priced paperback edition in the Wiley Classics Library.
†Now available in a lower priced paperback edition in the Wiley–Interscience Paperback Series.

HOCHBERG and TAMHANE · Multiple Comparison Procedures
HOCKING · Methods and Applications of Linear Models: Regression and the Analysis of Variance, *Third Edition*
HOEL · Introduction to Mathematical Statistics, *Fifth Edition*
HOGG and KLUGMAN · Loss Distributions
HOLLANDER, WOLFE, and CHICKEN · Nonparametric Statistical Methods, *Third Edition*
HOSMER and LEMESHOW · Applied Logistic Regression, *Second Edition*
HOSMER, LEMESHOW, and MAY · Applied Survival Analysis: Regression Modeling of Time-to-Event Data, *Second Edition*
HUBER · Data Analysis: What Can Be Learned From the Past 50 Years
HUBER · Robust Statistics
† HUBER and RONCHETTI · Robust Statistics, *Second Edition*
HUBERTY · Applied Discriminant Analysis, *Second Edition*
HUBERTY and OLEJNIK · Applied MANOVA and Discriminant Analysis, *Second Edition*
HUITEMA · The Analysis of Covariance and Alternatives: Statistical Methods for Experiments, Quasi-Experiments, and Single-Case Studies, *Second Edition*
HUNT and KENNEDY · Financial Derivatives in Theory and Practice, *Revised Edition*
HURD and MIAMEE · Periodically Correlated Random Sequences: Spectral Theory and Practice
HUSKOVA, BERAN, and DUPAC · Collected Works of Jaroslav Hajek— with Commentary
HUZURBAZAR · Flowgraph Models for Multistate Time-to-Event Data
JACKMAN · Bayesian Analysis for the Social Sciences
† JACKSON · A User's Guide to Principle Components
JOHN · Statistical Methods in Engineering and Quality Assurance
JOHNSON · Multivariate Statistical Simulation
JOHNSON and BALAKRISHNAN · Advances in the Theory and Practice of Statistics: A Volume in Honor of Samuel Kotz
JOHNSON, KEMP, and KOTZ · Univariate Discrete Distributions, *Third Edition*
JOHNSON and KOTZ (editors) · Leading Personalities in Statistical Sciences: From the Seventeenth Century to the Present
JOHNSON, KOTZ, and BALAKRISHNAN · Continuous Univariate Distributions, Volume 1, *Second Edition*
JOHNSON, KOTZ, and BALAKRISHNAN · Continuous Univariate Distributions, Volume 2, *Second Edition*
JOHNSON, KOTZ, and BALAKRISHNAN · Discrete Multivariate Distributions
JUDGE, GRIFFITHS, HILL, LÜTKEPOHL, and LEE · The Theory and Practice of Econometrics, *Second Edition*
JUREK and MASON · Operator-Limit Distributions in Probability Theory
KADANE · Bayesian Methods and Ethics in a Clinical Trial Design
KADANE AND SCHUM · A Probabilistic Analysis of the Sacco and Vanzetti Evidence
KALBFLEISCH and PRENTICE · The Statistical Analysis of Failure Time Data, *Second Edition*
KARIYA and KURATA · Generalized Least Squares
KASS and VOS · Geometrical Foundations of Asymptotic Inference
† KAUFMAN and ROUSSEEUW · Finding Groups in Data: An Introduction to Cluster Analysis
KEDEM and FOKIANOS · Regression Models for Time Series Analysis
KENDALL, BARDEN, CARNE, and LE · Shape and Shape Theory
KHURI · Advanced Calculus with Applications in Statistics, *Second Edition*
KHURI, MATHEW, and SINHA · Statistical Tests for Mixed Linear Models
* KISH · Statistical Design for Research

*Now available in a lower priced paperback edition in the Wiley Classics Library.
†Now available in a lower priced paperback edition in the Wiley–Interscience Paperback Series.

KLEIBER and KOTZ · Statistical Size Distributions in Economics and Actuarial Sciences
KLEMELÄ · Smoothing of Multivariate Data: Density Estimation and Visualization
KLUGMAN, PANJER, and WILLMOT · Loss Models: From Data to Decisions, *Third Edition*
KLUGMAN, PANJER, and WILLMOT · Loss Models: Further Topics
KLUGMAN, PANJER, and WILLMOT · Solutions Manual to Accompany Loss Models: From Data to Decisions, *Third Edition*
KOSKI and NOBLE · Bayesian Networks: An Introduction
KOTZ, BALAKRISHNAN, and JOHNSON · Continuous Multivariate Distributions, Volume 1, *Second Edition*
KOTZ and JOHNSON (editors) · Encyclopedia of Statistical Sciences: Volumes 1 to 9 with Index
KOTZ and JOHNSON (editors) · Encyclopedia of Statistical Sciences: Supplement Volume
KOTZ, READ, and BANKS (editors) · Encyclopedia of Statistical Sciences: Update Volume 1
KOTZ, READ, and BANKS (editors) · Encyclopedia of Statistical Sciences: Update Volume 2
KOWALSKI and TU · Modern Applied U-Statistics
KRISHNAMOORTHY and MATHEW · Statistical Tolerance Regions: Theory, Applications, and Computation
KROESE, TAIMRE, and BOTEV · Handbook of Monte Carlo Methods
KROONENBERG · Applied Multiway Data Analysis
KULINSKAYA, MORGENTHALER, and STAUDTE · Meta Analysis: A Guide to Calibrating and Combining Statistical Evidence
KULKARNI and HARMAN · An Elementary Introduction to Statistical Learning Theory
KUROWICKA and COOKE · Uncertainty Analysis with High Dimensional Dependence Modelling
KVAM and VIDAKOVIC · Nonparametric Statistics with Applications to Science and Engineering
LACHIN · Biostatistical Methods: The Assessment of Relative Risks, *Second Edition*
LAD · Operational Subjective Statistical Methods: A Mathematical, Philosophical, and Historical Introduction
LAMPERTI · Probability: A Survey of the Mathematical Theory, *Second Edition*
LAWLESS · Statistical Models and Methods for Lifetime Data, *Second Edition*
LAWSON · Statistical Methods in Spatial Epidemiology, *Second Edition*
LE · Applied Categorical Data Analysis, *Second Edition*
LE · Applied Survival Analysis
LEE · Structural Equation Modeling: A Bayesian Approach
LEE and WANG · Statistical Methods for Survival Data Analysis, *Fourth Edition*
LePAGE and BILLARD · Exploring the Limits of Bootstrap
LESSLER and KALSBEEK · Nonsampling Errors in Surveys
LEYLAND and GOLDSTEIN (editors) · Multilevel Modelling of Health Statistics
LIAO · Statistical Group Comparison
LIN · Introductory Stochastic Analysis for Finance and Insurance
LITTLE and RUBIN · Statistical Analysis with Missing Data, *Second Edition*
LLOYD · The Statistical Analysis of Categorical Data
LOWEN and TEICH · Fractal-Based Point Processes
MAGNUS and NEUDECKER · Matrix Differential Calculus with Applications in Statistics and Econometrics, *Revised Edition*
MALLER and ZHOU · Survival Analysis with Long Term Survivors
MARCHETTE · Random Graphs for Statistical Pattern Recognition
MARDIA and JUPP · Directional Statistics

*Now available in a lower priced paperback edition in the Wiley Classics Library.
†Now available in a lower priced paperback edition in the Wiley–Interscience Paperback Series.

MARKOVICH · Nonparametric Analysis of Univariate Heavy-Tailed Data: Research and Practice
MARONNA, MARTIN and YOHAI · Robust Statistics: Theory and Methods
MASON, GUNST, and HESS · Statistical Design and Analysis of Experiments with Applications to Engineering and Science, *Second Edition*
McCULLOCH, SEARLE, and NEUHAUS · Generalized, Linear, and Mixed Models, *Second Edition*
McFADDEN · Management of Data in Clinical Trials, *Second Edition*
* McLACHLAN · Discriminant Analysis and Statistical Pattern Recognition
McLACHLAN, DO, and AMBROISE · Analyzing Microarray Gene Expression Data
McLACHLAN and KRISHNAN · The EM Algorithm and Extensions, *Second Edition*
McLACHLAN and PEEL · Finite Mixture Models
McNEIL · Epidemiological Research Methods
MEEKER and ESCOBAR · Statistical Methods for Reliability Data
MEERSCHAERT and SCHEFFLER · Limit Distributions for Sums of Independent Random Vectors: Heavy Tails in Theory and Practice
MENGERSEN, ROBERT, and TITTERINGTON · Mixtures: Estimation and Applications
MICKEY, DUNN, and CLARK · Applied Statistics: Analysis of Variance and Regression, *Third Edition*
* MILLER · Survival Analysis, *Second Edition*
MONTGOMERY, JENNINGS, and KULAHCI · Introduction to Time Series Analysis and Forecasting
MONTGOMERY, PECK, and VINING · Introduction to Linear Regression Analysis, *Fifth Edition*
MORGENTHALER and TUKEY · Configural Polysampling: A Route to Practical Robustness
MUIRHEAD · Aspects of Multivariate Statistical Theory
MULLER and STOYAN · Comparison Methods for Stochastic Models and Risks
MURTHY, XIE, and JIANG · Weibull Models
MYERS, MONTGOMERY, and ANDERSON-COOK · Response Surface Methodology: Process and Product Optimization Using Designed Experiments, *Third Edition*
MYERS, MONTGOMERY, VINING, and ROBINSON · Generalized Linear Models. With Applications in Engineering and the Sciences, *Second Edition*
NATVIG · Multistate Systems Reliability Theory With Applications
† NELSON · Accelerated Testing, Statistical Models, Test Plans, and Data Analyses
† NELSON · Applied Life Data Analysis
NEWMAN · Biostatistical Methods in Epidemiology
NG, TAIN, and TANG · Dirichlet Theory: Theory, Methods and Applications
OKABE, BOOTS, SUGIHARA, and CHIU · Spatial Tesselations: Concepts and Applications of Voronoi Diagrams, *Second Edition*
OLIVER and SMITH · Influence Diagrams, Belief Nets and Decision Analysis
PALTA · Quantitative Methods in Population Health: Extensions of Ordinary Regressions
PANJER · Operational Risk: Modeling and Analytics
PANKRATZ · Forecasting with Dynamic Regression Models
PANKRATZ · Forecasting with Univariate Box-Jenkins Models: Concepts and Cases
PARDOUX · Markov Processes and Applications: Algorithms, Networks, Genome and Finance
PARMIGIANI and INOUE · Decision Theory: Principles and Approaches
* PARZEN · Modern Probability Theory and Its Applications
PEÑA, TIAO, and TSAY · A Course in Time Series Analysis
PESARIN and SALMASO · Permutation Tests for Complex Data: Applications and Software

*Now available in a lower priced paperback edition in the Wiley Classics Library.
†Now available in a lower priced paperback edition in the Wiley–Interscience Paperback Series.

PIANTADOSI · Clinical Trials: A Methodologic Perspective, *Second Edition*
POURAHMADI · Foundations of Time Series Analysis and Prediction Theory
POURAHMADI · High-Dimensional Covariance Estimation
POWELL · Approximate Dynamic Programming: Solving the Curses of Dimensionality, *Second Edition*
POWELL and RYZHOV · Optimal Learning
PRESS · Subjective and Objective Bayesian Statistics, *Second Edition*
PRESS and TANUR · The Subjectivity of Scientists and the Bayesian Approach
PURI, VILAPLANA, and WERTZ · New Perspectives in Theoretical and Applied Statistics
† PUTERMAN · Markov Decision Processes: Discrete Stochastic Dynamic Programming
QIU · Image Processing and Jump Regression Analysis
* RAO · Linear Statistical Inference and Its Applications, *Second Edition*
RAO · Statistical Inference for Fractional Diffusion Processes
RAUSAND and HØYLAND · System Reliability Theory: Models, Statistical Methods, and Applications, *Second Edition*
RAYNER, THAS, and BEST · Smooth Tests of Goodnes of Fit: Using R, *Second Edition*
RENCHER and SCHAALJE · Linear Models in Statistics, *Second Edition*
RENCHER and CHRISTENSEN · Methods of Multivariate Analysis, *Third Edition*
RENCHER · Multivariate Statistical Inference with Applications
RIGDON and BASU · Statistical Methods for the Reliability of Repairable Systems
* RIPLEY · Spatial Statistics
* RIPLEY · Stochastic Simulation
ROHATGI and SALEH · An Introduction to Probability and Statistics, *Second Edition*
ROLSKI, SCHMIDLI, SCHMIDT, and TEUGELS · Stochastic Processes for Insurance and Finance
ROSENBERGER and LACHIN · Randomization in Clinical Trials: Theory and Practice
ROSSI, ALLENBY, and McCULLOCH · Bayesian Statistics and Marketing
† ROUSSEEUW and LEROY · Robust Regression and Outlier Detection
ROYSTON and SAUERBREI · Multivariate Model Building: A Pragmatic Approach to Regression Analysis Based on Fractional Polynomials for Modeling Continuous Variables
* RUBIN · Multiple Imputation for Nonresponse in Surveys
RUBINSTEIN and KROESE · Simulation and the Monte Carlo Method, *Second Edition*
RUBINSTEIN and MELAMED · Modern Simulation and Modeling
RUBINSTEIN, RIDDER, and VAISMAN · Fast Sequential Monte Carlo Methods for Counting and Optimization
RYAN · Modern Engineering Statistics
RYAN · Modern Experimental Design
RYAN · Modern Regression Methods, *Second Edition*
RYAN · Sample Size Determination and Power
RYAN · Statistical Methods for Quality Improvement, *Third Edition*
SALEH · Theory of Preliminary Test and Stein-Type Estimation with Applications
SALTELLI, CHAN, and SCOTT (editors) · Sensitivity Analysis
SCHERER · Batch Effects and Noise in Microarray Experiments: Sources and Solutions
* SCHEFFE · The Analysis of Variance
SCHIMEK · Smoothing and Regression: Approaches, Computation, and Application
SCHOTT · Matrix Analysis for Statistics, *Second Edition*
SCHOUTENS · Levy Processes in Finance: Pricing Financial Derivatives
SCOTT · Multivariate Density Estimation: Theory, Practice, and Visualization
* SEARLE · Linear Models
† SEARLE · Linear Models for Unbalanced Data
† SEARLE · Matrix Algebra Useful for Statistics
† SEARLE, CASELLA, and McCULLOCH · Variance Components

*Now available in a lower priced paperback edition in the Wiley Classics Library.
†Now available in a lower priced paperback edition in the Wiley–Interscience Paperback Series.

SEARLE and WILLETT · Matrix Algebra for Applied Economics
SEBER · A Matrix Handbook For Statisticians
† SEBER · Multivariate Observations
SEBER and LEE · Linear Regression Analysis, *Second Edition*
† SEBER and WILD · Nonlinear Regression
SENNOTT · Stochastic Dynamic Programming and the Control of Queueing Systems
* SERFLING · Approximation Theorems of Mathematical Statistics
SHAFER and VOVK · Probability and Finance: It's Only a Game!
SHERMAN · Spatial Statistics and Spatio-Temporal Data: Covariance Functions and Directional Properties
SILVAPULLE and SEN · Constrained Statistical Inference: Inequality, Order, and Shape Restrictions
SINGPURWALLA · Reliability and Risk: A Bayesian Perspective
SMALL and McLEISH · Hilbert Space Methods in Probability and Statistical Inference
SRIVASTAVA · Methods of Multivariate Statistics
STAPLETON · Linear Statistical Models, *Second Edition*
STAPLETON · Models for Probability and Statistical Inference: Theory and Applications
STAUDTE and SHEATHER · Robust Estimation and Testing
STOYAN · Counterexamples in Probability, *Second Edition*
STOYAN, KENDALL, and MECKE · Stochastic Geometry and Its Applications, *Second Edition*
STOYAN and STOYAN · Fractals, Random Shapes and Point Fields: Methods of Geometrical Statistics
STREET and BURGESS · The Construction of Optimal Stated Choice Experiments: Theory and Methods
STYAN · The Collected Papers of T. W. Anderson: 1943–1985
SUTTON, ABRAMS, JONES, SHELDON, and SONG · Methods for Meta-Analysis in Medical Research
TAKEZAWA · Introduction to Nonparametric Regression
TAMHANE · Statistical Analysis of Designed Experiments: Theory and Applications
TANAKA · Time Series Analysis: Nonstationary and Noninvertible Distribution Theory
THOMPSON · Empirical Model Building: Data, Models, and Reality, *Second Edition*
THOMPSON · Sampling, *Third Edition*
THOMPSON · Simulation: A Modeler's Approach
THOMPSON and SEBER · Adaptive Sampling
THOMPSON, WILLIAMS, and FINDLAY · Models for Investors in Real World Markets
TIERNEY · LISP-STAT: An Object-Oriented Environment for Statistical Computing and Dynamic Graphics
TSAY · Analysis of Financial Time Series, *Third Edition*
TSAY · An Introduction to Analysis of Financial Data with R
UPTON and FINGLETON · Spatial Data Analysis by Example, Volume II: Categorical and Directional Data
† VAN BELLE · Statistical Rules of Thumb, *Second Edition*
VAN BELLE, FISHER, HEAGERTY, and LUMLEY · Biostatistics: A Methodology for the Health Sciences, *Second Edition*
VESTRUP · The Theory of Measures and Integration
VIDAKOVIC · Statistical Modeling by Wavelets
VIERTL · Statistical Methods for Fuzzy Data
VINOD and REAGLE · Preparing for the Worst: Incorporating Downside Risk in Stock Market Investments
WALLER and GOTWAY · Applied Spatial Statistics for Public Health Data
WEISBERG · Applied Linear Regression, *Third Edition*
WEISBERG · Bias and Causation: Models and Judgment for Valid Comparisons

*Now available in a lower priced paperback edition in the Wiley Classics Library.
†Now available in a lower priced paperback edition in the Wiley–Interscience Paperback Series.

WELSH · Aspects of Statistical Inference
WESTFALL and YOUNG · Resampling-Based Multiple Testing: Examples and Methods for *p*-Value Adjustment
* WHITTAKER · Graphical Models in Applied Multivariate Statistics
WINKER · Optimization Heuristics in Economics: Applications of Threshold Accepting
WOODWORTH · Biostatistics: A Bayesian Introduction
WOOLSON and CLARKE · Statistical Methods for the Analysis of Biomedical Data, *Second Edition*
WU and HAMADA · Experiments: Planning, Analysis, and Parameter Design Optimization, *Second Edition*
WU and ZHANG · Nonparametric Regression Methods for Longitudinal Data Analysis
YIN · Clinical Trial Design: Bayesian and Frequentist Adaptive Methods
YOUNG, VALERO-MORA, and FRIENDLY · Visual Statistics: Seeing Data with Dynamic Interactive Graphics
ZACKS · Stage-Wise Adaptive Designs
* ZELLNER · An Introduction to Bayesian Inference in Econometrics
ZELTERMAN · Discrete Distributions—Applications in the Health Sciences
ZHOU, OBUCHOWSKI, and McCLISH · Statistical Methods in Diagnostic Medicine, *Second Edition*

*Now available in a lower priced paperback edition in the Wiley Classics Library.
†Now available in a lower priced paperback edition in the Wiley–Interscience Paperback Series.